Spin Dynamics and Damping in Ferromagnetic Thin Films and Nanostructures

Anjan Barman · Jaivardhan Sinha

Spin Dynamics and Damping in Ferromagnetic Thin Films and Nanostructures

 Springer

Anjan Barman
Department of Condensed Matter Physics
  and Material Sciences
S. N. Bose National Centre for Basic
  Sciences
Kolkata
India

Jaivardhan Sinha
Department of Condensed Matter Physics
  and Material Sciences
S. N. Bose National Centre for Basic
  Sciences
Kolkata
India

ISBN 978-3-319-88213-0   ISBN 978-3-319-66296-1   (eBook)
https://doi.org/10.1007/978-3-319-66296-1

Printed on acid-free paper

This Springer imprint is published by Springer Nature
The registered company is Springer International Publishing AG
The registered company address is: Gewerbestrasse 11, 6330 Cham, Switzerland

Preface

This book presents the recent development in the area of spin dynamics and magnetic damping. It is important to understand the basic physical phenomena that govern the spin dynamics in the ferromagnetic thin films and nanostructures along with its application potential in data storage, memory, and communication technology. One of the crucial parameters that govern the spin dynamics is magnetic damping. The book begins with introducing spin dynamics with a perspective of quantum mechanical approach. The phenomenological Landau–Lifshitz–Gilbert (LLG) equation describing magnetization dynamics and also where the Gilbert damping appears is introduced subsequently. Discussion about the origin and mechanism (e.g., intrinsic and extrinsic, local and non-local) of damping is included. Existing theoretical models, e.g., breathing Fermi surface model, s-d-exchange relaxation model, spin-flip scattering along with their limitations in describing recent results are covered in reasonable detail. These descriptions are included for broad range of readers including graduate students. We have discussed the important experimental techniques for investigating magnetization dynamics starting from elementary level. Effect of material parameter, and electrical and optical control of magnetization dynamics are discussed along with recent development in this field of research.

The main highlights of the book are new experimental approaches for controlling damping in ferromagnetic thin films and nanostructures. The mechanism for the damping control in metallic ferromagnet/non-magnetic metal bilayer film is given emphasis keeping in mind its usefulness in spintronics-based devices. Various ways to tune the damping, specifically, dynamic (using spin current from spin Hall effect) and static control (interface intermixing) in ferromagnetic layer/heavy metal layer are described. For the investigation of modulation of damping, all-optical detection techniques, for example, time-resolved magneto-optical Kerr effect microscope and Brillouin light scattering technique, are discussed, in particular giving emphasis to the advantages of implementing them. Specifically, experimental results for $Pt/Ni_{81}Fe_{19}$ bilayer stack have been described in detail. A new method for estimation of spin Hall angle using all-optical detection technique is shown to be more appropriate as it excludes several sources of error that are present in electrical detection techniques. Furthermore, the results of large spin Hall angle material in achieving larger modulation of damping are presented. To best of our knowledge, these results have been published quite recently in scientific journals and are described for the first time in such great detail to become part of a book. Invoking the concept of dynamic and static control of damping is another unique feature of this book. We thus hope that this book provides up-to-date overview of spin dynamics and magnetic damping along with future outlook which will be useful for graduate students as well as advanced researchers working in this field.

Kolkata, India   Anjan Barman
                                                            Jaivardhan Sinha

Acknowledgements

We gratefully acknowledge the financial support from Department of Science and Technology, Government of India, under Grant No. SR/NM/NS-09/2011(G) and DST/INT/UK/P-44/2012. We thank Chandrima Banerjee for her help in preparing the description of Brillouin light scattering. We thank Bivas Rana, Semanti Pal, Susmita Saha, Arnab Ganguly, Samiran Choudhury, and Sucheta Mondal for their contribution in original research which became part of the book. We are also indebted to our colleagues involved in collaborative research works which have been discussed in this book.

Contents

Chapter 1
Introduction

Magnetism has been a subject of intense research for more than a century as it has played a pivotal role for the benefit of mankind. Starting from old-age magnetic compass used for navigation, magnetism has numerous applications in modern and future data storage technology. Magnetism research has initiated several important fundamental phenomena and took an important role in modern science and technology. An extremely interesting aspect of magnetism research is that the gap between the theoretical predictions of the phenomena and the application is relatively small. Magnetism has been characterized in few classes depending on the type of response of the material in the presence of magnetic field. These are diamagnetism, paramagnetism, ferromagnetism, ferrimagnetism, and antiferromagnetism. In this book, we will discuss mainly ferromagnetism.

In general, many aspects of ferromagnetism were attempted to understand in terms of their classical analogs. The basis of magnetostatics and magnetodynamics was mathematically formulated by Maxwell in 1862 which was based on early research works of Gauss, Faraday, and Lorentz. Known as Maxwell's [1] equation, a unified theory of electricity, magnetism, and optics was postulated by him. Despite the Maxwell's proposal, there were numerous unexplained issues while trying to invoke classical mechanics to explain magnetism [2, 3]. Historically, the first experiment related to the observation of the imprint of quantum nature of spins was performed by Otto Stern and Walther Gerlac in 1922 [4]. In their experiment, mainly silver atoms were used in the presence of inhomogeneous magnetic field to study such effect. However, the case of electrons is non-trivial and the Lorentz force acting on it combined with uncertainty principle prevents the separation of opposite spins. Interestingly, the discovery by Uhlenbeck and Goudsmit [5] in 1925 that the Amperian current is associated with quantized angular momentum, and especially with the intrinsic spin of the electron, led to an in-depth understanding of the relation between electricity and magnetism. It was established that the spin is quantized in such a way that it can have just two possible orientations in a magnetic field, 'up' and 'down.' The source of the intrinsic magnetic moment of the electron is spin and is defined as the Bohr magneton: $\mu_B = 9.274 \times 10^{-24}$ A/m^2.

© Springer International Publishing AG 2018
A. Barman and J. Sinha, *Spin Dynamics and Damping in Ferromagnetic Thin Films and Nanostructures*, https://doi.org/10.1007/978-3-319-66296-1_1

Essentially, the magnetic properties of solids arise from the magnetic moments of the atomic electrons. In 1929, Heisenberg [6] showed by taking into account Pauli principle that the interaction responsible for ferromagnetism can be considered as electrostatic in nature along with exchange. It became important to invoke quantum mechanics to explain the phenomena such as the quantization of angular momentum and certain interactions between spins. The global electronic spin alignment in the material is the origin of magnetic moment. It mainly arises if parallel spin alignment with respect to the neighboring atomic spins is energetically favorable. In a broad sense, it can be considered that the interaction of spins in the ferromagnetic material results in the net magnetization. On the microscopic scale where all spins are aligned, the strength of the magnetic moment and thus the net magnetization (M defined as magnetic moment per unit volume) are constant. Hence, one can change the direction of M by some external means, for example, by an external magnetic field with which the magnetization tries to align. However, the crystal structure and shape of a ferromagnetic specimen also play a decisive role in determining the magnetization direction. Overall, in a ferromagnetic specimen the magnetization orientation is described in terms of easy and hard axis with the magnetization trying to align preferentially along the easy axis. An important point to mention here is that we will discuss magnetization dynamics mainly in thin films and confined structures in this book [7, 8].

Spin dynamics or commonly referred as magnetization dynamics in ferromagnetic thin films and confined structures is one of the most fascinating phenomena in magnetism which has drawn significant attention of researchers working in the fundamental aspects of magnetism as well as applications. Understanding of the spin dynamics is essential for developing spintronics-based devices [9]. Briefly, just as conventional electronic devices utilize the charge degree of freedom of electron, spin-based electronic (spintronic) devices utilize spin degree of freedom. The flow of spin, like the movement of charge, can be used as information carriers among devices. The advantage of utilizing the flow of spin over charge is that spin can be easily tuned by externally applied tiny magnetic fields. Interestingly, more significant property of spin is its long coherence or relaxation time. Due to this, the created spin tends to preserve its state for a long time, unlike charge states, which are easily perturbed by scattering with defects and impurities. The spin dynamics refers to the dynamics of either the population and the phase of the spin of an ensemble of particles, or a coherent spin manipulation of a single- or a few-spin system [9]. In order to explain magnetization dynamics, in 1935, Landau and Lifshitz [10] proposed an equation which describes the magnetization precession. Gilbert [11] introduced a phenomenological damping term in the Landau–Lifshitz equation to practically illustrate the dynamical magnetization precession and subsequent alignment of magnetization along the effective field direction. Interesting phenomena, namely ferromagnetic resonance (FMR), was experimentally observed by Griffiths in 1946 [12]. Kittel [13], in 1948, derived the expression for the FMR frequency by taking into account the external magnetic field and internal magnetic parameters. In this book, we provide an overview of spin dynamics along with relevant details. The structure of this book is described below.

Chapter 2 of this book describes magnetization dynamics at various time scales. Spin dynamics is a rich topic, and it can occur over a wide range of time scale. Depending on the associated characteristic time scales involved with the spin dynamics, it can be categorized into various types. Theoretically, it is known that the time scales are determined by the interaction energies via Heisenberg relation. Up to date, the known fastest process is the fundamental exchange interaction and its time scale is within 10 fs. The spin–orbit coupling and spin-transfer torque have rather broader time scale ~ 10 fs–1 ps. Experimentally, an elegant way to study the ultrafast spin dynamics is to excite the ferromagnetic material using ultrashort laser pulses [14]. In such studies, it is established that the time scale for laser-induced ultrafast demagnetization is of the order of few hundreds of fs. Furthermore, the time scale of 1–10 ps is defined for the fast remagnetization time following the ultrafast demagnetization. Given these short time scales involved in various magnetization manipulation processes, these become the most preferred choice for magnetic data storage industry where there is an ever-growing demand of writing and reading data at smallest possible time scales. In the present-day technology, the magnetic writing is done via reversal of spin and it has a time scale of few ns to few hundreds of ns. Furthermore, the precessional magnetization dynamics has the time scale of the order of few ps to few hundreds of ps and it is followed by the damping of precession over a time scale for few ns as schematically depicted in Fig. 1.1. The intriguing topic of optically induced ultrafast spin dynamics has been central to address numerous fundamental aspects. Though this topic is investigated over a decade, still some of the interesting experimental observations are not well understood. To this end, in the first pioneering work from the group of Beaurepaire et al. [15] in 1996, it was found that a nickel thin film can be demagnetized in a sub-ps time scale after excitation with a sub-100 fs pulsed laser beam.

Following this, there were several studies on various magnetic metals and alloys which confirmed the above observation. It triggered several theoretical studies as

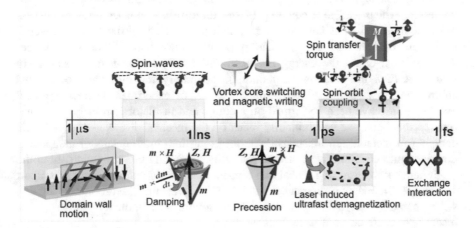

Fig. 1.1 Time scales of various dynamical magnetic processes

well, and some of it have been partially successful in explaining the results; however, in-depth insight of 'how the magnetic moment gets quenched at such short time scale and angular momentum still remain conserved' is missing [16, 17]. For more detailed description of ultrafast demagnetization, readers may follow the available literature [14, 18, 19]. In this book, we discuss mainly the processes involved in spin dynamics beyond ultrafast demagnetization. It is known that following the demagnetization, the electronic charges and spins start to relax. Two different time scales ranging between few ps to few ns are known to be associated with magnetization relaxation. Overall, the magnetization relaxation in these systems follows bi-exponential decay. Among these two different time scales, the faster one contains rich information about the relaxation time of the hot electrons and spins. Electron–phonon interactions mediate the energy exchange process between hot electrons, spins and lattice. This fast relaxation time may vary from few ps to tens of ps, and it is discussed in several references. It is understood that during the fast relaxation time, by exchanging heat, spins come to an equilibrium temperature with charge and lattice which effectively changes the lattice temperature. The change in the lattice temperature leads to change in magnetocrystalline anisotropy. Interestingly, the ultrafast change in magnetocrystalline anisotropy acts as an effective field pulse and triggers the precession of magnetization. The longer relaxation time (sub-ns–few ns) of the bi-exponential decay corresponds to the diffusion of electron and lattice heat to the surroundings (such as substrate in case of thin films). During this time scale, the precession of magnetization damps out and practically all the dynamical processes are ceased.

In Chap. 3 of this book, we describe the magnetic damping in great detail. Although the processes of energy dissipation in ferromagnets can be complicated, it is approximately described by a single damping parameter. For practical systems, Gilbert [11] proposed a phenomenological term used to describe the damping of the magnetization precession. Depending on the time derivative of the magnetization, Gilbert modeled a 'viscous' damping by considering a phenomenological dissipation term. Actually, the damping parameter contains contributions from a variety of processes, which lead to a coupling of the uniform and non-uniform precession modes and other thermal baths of the magnetically ordered substance. Thus, overall in this process, the dissipation of energy occurs to non-magnetic degrees of freedom. In most cases, the energy is transferred finally to the lattice resulting in heating it (i.e., the creation of phonons). To clarify the different mechanisms contributing to damping, a considerable amount of experimental and theoretical investigations have been performed on metallic and insulator ferromagnetic materials. In this book, we will primarily focus on the mechanisms related to magnetic damping in metallic ferromagnets. Without going into more details of magnetization dynamics equation, we concentrate in this chapter on the discussion about the origin and mechanism (e.g., local and non-local, intrinsic and extrinsic) of Gilbert damping [20]. Existing theoretical models (e.g., breathing Fermi surface model, s-d exchange model, spin-flip scattering) along with their limitations in describing recent results are covered in reasonable detail.

Dedicated effort to investigate the magnetization dynamics experimentally with deep theoretical understanding was made by various research groups in the last few decades. Different techniques in the frequency, wave vector, and time domains were developed to probe the finer details of magnetization dynamics. The conventional ferromagnetic resonance (FMR) is a frequency domain technique, where the sample is excited at a particular frequency. The external bias field is swept to probe the magnetization dynamics through the resonance. Another efficient technique is Brillouin light scattering (BLS) technique which has emerged as a quite efficient method to measure the magnetization dynamics in the wave vector domain [21]. Recent developments of space-resolved and time-resolved BLS techniques have taken the application of BLS to much advanced level. The time-resolved magneto-optical Kerr effect (TR-MOKE) microscopy offers high time resolution along with a spatial resolution in the sub-μm regime [19]. With the time resolution of the order of subhundred femtosecond (limited by the pulse width of the laser), the ultrafast magnetization dynamics can be probed quite efficiently. Further by incorporating a scanning microscope with the TR-MOKE, time-resolved scanning Kerr microscopy (TR-SKM) was developed to image the time evolution of spatial distribution of magnetization in confined magnetic elements. In Chap. 4 of this book, we provide an overview of some of these interesting experimental setups, their working principles, and advantages and disadvantages.

Chapter 5 of this book is related to understanding of the factors that influence magnetization dynamics. In the context of developing general understanding of the spin dynamics, two open problems need to be addressed. One of those is related to the influence of the particular sample which is under study and other is related to the accuracy and interpretation of the results, obtained with a given experimental technique, in terms of spin dynamics. In order to exploit the full potential of spintronics application, ongoing cutting-edge research for developing ferromagnetic thin films, multilayers, and nanostructures is pursued. For the case of sample, specifically while using ferromagnetic thin films, multilayers, and nanostructures, the role of shape, crystallinity, and interface becomes important [8, 19, 22–24]. In this chapter of this book, we discuss the aspects of magnetization dynamics with these two questions in mind. In particular, emphasis is given on emergent magnetic properties at the interface in ferromagnetic thin-film heterostructures such as the interfacial intermixing and spin Hall effect. For more established interfacial phenomena such as exchange bias and exchange spring magnets, interested readers may follow other existing literature. We categorize the intrinsic and extrinsic interfacial magnetic phenomena in the context of its influence on magnetization dynamics. Briefly, we consider defects, interdiffusion, interfacial roughness produced by structural and/or chemical disorder.

We believe that the description provided up to Chap. 5 in this book gives flavor of the significance of control of magnetization dynamics. We mainly discuss the well-established basic mechanisms up to Chap. 5. In the remaining chapters of this book, we intend to provide the recent advances in this field with a particular emphasis on electrical and optical control of spin dynamics. The need of an in-depth understanding of the electrical and optical control of spin dynamics arises primarily due to the ever-increasing demand of information processing in the magnetic recording industry. Of course, the rich fundamental physics involved with it has drawn the interest of the researchers. A fundamental phenomenon relating to charge current and magnetization is the anisotropic magnetoresistance (AMR). The electrical resistivity of a magnetic system that depends on the relative angle between the applied charge current and magnetization direction is known as anisotropic magnetoresistance. In 1988, the phenomenon of giant magnetoresistance (GMR) was discovered by Grünberg and Fert in Fe/Cr superlattices [25, 26]. It was found that the two adjacent ferromagnetic layers of the system separated by a non-magnetic spacer are usually antiferromagnetically coupled to each other by an indirect exchange interaction in turn leading to large electrical resistance of the system. In the presence of an external magnetic field, the moments of the magnetic layers align themselves in the same direction resulting in a significant drop in the electrical resistance of the system. This phenomenon has enormous application in the present-day magnetic recording industry, and it is considered as one of the most remarkable discoveries in the field of spintronics to be awarded Noble Prize in 2007. For exciting the spin dynamics using electrical current, one of the prerequisites is to obtain spin-polarized current. We begin Chap. 6 with the discussion of methods to achieve spin-polarized current and its application for generating spin-transfer torque (STT). The mechanism of spin-transfer torque given by Sclonzewski and Berger independently in 1996 added a new dimension to modern spintronic research [27, 28]. Their findings suggested that when a spin-polarized electrical current interacts with a ferromagnet, a portion of the spin angular momentum carried by the electrons can be transferred to the magnet. As the time rate of change of angular momentum is defined as torque, so it was concluded that the spin-polarized electrons may apply a torque directly to the magnet [29, 30]. Based on this concept, in 1998, Tsoi et al. [31] experimentally demonstrated magnetization precession in Co/Cu multilayers by currents injected by a point contact. In 1999, Myers et al. observed switching in the orientation of magnetic moments in Co/Cu/Co sandwich structures by currents perpendicular to plane. Experimentally, it has been demonstrated that in nanomagnetic devices of size ∼ 200 nm, this STT can be much stronger per unit current than the torque on the magnet due to the magnetic field that is generated by the current. Thus, the STT is of interest as an alternative and potentially more efficient mechanism for triggering magnetization dynamics using charge current. After the discovery of giant magnetoresistance, for about a decade, primarily all spintronic concepts were based on the idea of generating a spin-polarized current by passing a charge current through a ferromagnetic layer and utilizing these spin-polarized current for triggering magnetization dynamics. More recently, it has been realized that even through non-magnetic heavy metals the large spin–orbit interactions can provide an efficient alternative pathway for generating spin currents when the charge current flows through it. Theoretically, Dyakonov and Perel [32] predicted in 1971 this key phenomenon known as spin Hall effect. To describe in brief, an initially unpolarized charge current can be converted into a transverse spin current either by the extrinsic spin-dependent scattering or by the intrinsic spin-dependent transverse velocities determined by the electronic band structure, which leads to spin accumulation at the transverse edges of the heavy metals. This spin current has been used to excite magnetization dynamics in a ferromagnet attached to the non-magnet in a non-magnet/ferromagnet heterostructure by the spin-transfer torque. Furthermore, the concept of spin Hall effect can be considered under the ultimate limit by passing the charge current through a two-dimensional system, which can generate a current-induced spin accumulation. Theoretically, such current-induced spin accumulation has been reported for the case of Rashba spin–orbit coupling [33]. Moreover, experimentally it has been shown recently that Rashba states at metallic interfaces can result in significant spin-transfer torques [34]. It is worth mentioning that the time scales involved in electrical control of magnetization dynamics is generally of the order of nanoseconds. Today's magnetic bits primarily respond in nanoseconds time scales; however, there is possibility and room of nearly 1000 times improvements if spin-photonic interaction can be utilized for magnetization switching.

Apart from the above-mentioned electrical control of magnetization dynamics, great interest and effort has been devoted by researchers to understand the optical control of magnetization dynamics. Interestingly, Kirilyuk et al. [14] found that using circularly polarized ultrafast laser pulses, magnetization switching can be reproducibly achieved in the compounds of rare earth–transition metal ferrimagnetic films. Subsequent theoretical works proposed two different mechanisms for understanding the experimental results. One of those involved the transfer of angular momentum from the circularly polarized light pulse to the spin system of the medium, and another invokes the action of spin currents induced by the laser. These intriguing ideas helped in partially understanding the interaction of light and matter on ultrashort time scales. In 2014, the work of Mangin et al. [35] demonstrated that a broad range of materials including synthetic ferrimagnetic multilayer films can be optically switched via ultrafast laser pulses. In the follow-up work, Lambert et al. [36] further generalized the materials systems including ferromagnetic Co/Pt multilayer films that could be optically switched using linearly polarized light (that possesses no net angular momentum) as well as left and right circularly polarized light. These recent research results hint at the possibility of near future technological application of optically induced magnetization switching in magnetic recording industry.

Generally, the magnetic recording media such as tape or disc incorporates a thin film of magnetic material on a rigid or flexible substrate. Sophisticated magnetic multilayer structures used in read/write heads for magnetic recording, magnetic sensors and magnetic random-access memory, are considered as the first-generation products of the spintronics age. The medium or a device which is composed of non-magnetic as well as magnetic materials is even more interesting as the properties of magnetic materials can be tuned by the choice of neighboring non-magnetic metal and the interface between non-magnet and ferromagnetic material. In Chap. 7 of this book, we present the results of static control of magnetic damping which is extremely important to control the magnetization dynamics in non-magnet/ferromagnet bilayer system. Damping plays a crucial role in spin-transfer torque magnetoresistive random-access memory (STT-MRAM) devices which are considered to be the new-generation spintronics-based devices [37]. Another emerging technology that significantly depends on damping is magnonics [38]. The ultimate goal of magnonics is to use magnons (spin-wave excitations) instead of diffusive spin current for encoding and processing information. In the magnonic device, the spin-wave propagation is limited by the material-dependent Gilbert damping. Control of magnetic damping at micro-nanoscale in non-magnet/ferromagnet bilayer system is desired for technological application in both spintronic and magnonic devices. The low damping is known to facilitate a lower writing current in STT-MRAM and longer propagation of spin waves in magnonic devices. Higher damping is desirable to increase the reversal rates and the coherent reversal of magnetic elements, as damping suppresses the precessional motion of the magnetization vector. Damping in such bilayer systems may get enhanced due to several mechanisms. Spin–orbit coupling (SOC) and interfacial d-d hybridization (intrinsic contribution) result in significantly enhanced damping. For in-plane magnetized thin films, another major contributor of large damping value is two-magnon scattering (extrinsic contribution) which is related to roughness and defects of the interface layer. Two-magnon scattering refers to the scattering of uniform magnetization precession into pairs of magnons with nonzero wave vectors. It may occur when the symmetry of the system is disturbed by structural defects like film roughness and intermixing. The total damping is the sum of intrinsic and extrinsic contributions [20]. By considering intrinsic and extrinsic contribution to damping, we discuss the results of ferromagnet/non-magnet (e.g., $Ni_{81}Fe_{19}$/Pt and Co/Pt) bi-layer films with controlled variation of non-magnetic layer thickness [39]. Our investigations of these magnetic bilayers suggested that in these the interface plays a crucial role in controlling the magnetization dynamics. Furthermore, we discuss the significance of interfacial engineering in ferromagnetic systems which offers an exciting opportunity to explore fundamental aspects of the interface physics as well as to address the technological requirements for control of damping. A sophisticated experimental technique, namely Focussed Ion Beam (FIB), which uses low-dose Ga^{+} ion irradiation for direct nanoscale patterning, is used to controllably tune the $Ni_{81}Fe_{19}$/Pt, $Ni_{81}Fe_{19}$/Au, and $Ni_{81}Fe_{19}$/Cr interface and in turn the damping. This technique of systematic interface engineering allows the precise control of local magnetization dynamics in bilayer and multilayer system without causing substantial structural changes or damage [40, 41]. A custom-built time-resolved magneto-optical Kerr effect (TR-MOKE) microscope has been used to investigate the variation of the effective damping, precession frequency, and, importantly, the spatial coherence of the spin dynamics. It is worth mentioning here that the static control of damping achieved by interfacial intermixing is an irreversible process and given the ever-increasing demand of the advanced spintronic and magnonic devices, it is desirable to control the damping reversibly in a dynamic fashion. We present in Chap. 7 the experimental results for achieving dynamic control of magnetic damping in heavy metal/ferromagnet bilayer system by utilizing spin Hall effect that originates due to large spin–orbit coupling in heavy metal. As mentioned earlier, the spin current due to SHE affects the magnetization dynamics of ferro-magnetic layer placed next to the heavy metal layer. SHE applies a spin torque which is collinear with the damping torque, and the sign of it can be varied depending on the direction of the flow of charge current through the heavy metal layer. We thus apply charge current through the heavy metal layer and evaluate the SHE-induced effect on magnetization dynamics via estimating modulation of effective damping in TR-MOKE experiments. We have first demonstrated the proof of concept of this experiment on a conventional Pt/NiFe bilayer film and subse-quently investigated the detailed thickness dependence of SHE of W on W/CoFeB bilayer films [42, 43]. Through these experiments, we explicitly provide a better estimate of the spin Hall angle of Pt and W which is found to be slightly improved with respect to the values obtained from earlier existing literature. Interestingly, we find that the extent of modulation of effective damping is well correlated with the spin Hall angle. From the point of view of spintronics and magnonics device applications, the ability to dynamically and reversibly tune effective damping using external stimuli as charge current is an important step forward.

Finally, in Chap. 8, we present the summary and future direction based on the topics covered in this book. We have presented a summary of spin dynamics following quantum mechanical approach and importantly the factors affecting it. Particularly, we have emphasized on the magnetic damping parameter by consid-ering its origin as well as tunability. We describe a number of techniques involving spin current, interfacial changes of a ferromagnetic system, or the combination of the heavy metal/ferromagnet bilayer system for tuning magnetic properties. Specifically, the experimental results covering the effective damping, changes in precessional frequency, and ultrafast relaxation processes in heavy metal/ferromagnet bilayer are presented in reasonable detail. All these results related to local modification of magnetic properties open a new avenue for con-structing nanoscale printed spintronic circuit, where data storage, data transfer, and logical operation can be performed simultaneously on a single thin-film structure. Most importantly, in many cases, the thin-film bilayer investigated in the experi-mental works also contains an oxide capping layer in the stack. The presence of the third layer allows one to visualize these stacks as thin-film heterostructure with structure inversion asymmetry. While in this book we mainly dealt with exchange interactions in ferromagnetic thin film, which lead to collinear alignment of lattice spins, in materials with broken inversion symmetry and strong spin–orbit coupling, the Dzyaloshinskii–Moriya interaction (DMI) which is an antisymmetric exchange interaction can stabilize canted spins. Of particular interest in such materials are magnetic skyrmions which are topologically protected particle-like chiral spin textures. Magnetic skyrmions can arrange spontaneously into lattices, and charge currents can displace them at remarkably low current densities. It will be really interesting to explore rich optically induced ultrafast skyrmion dynamics and its correlation with the fundamental aspect of magnetism.

The whole idea of this book is to stimulate the interest of readers in variety of areas open in the field of spin dynamics.

References

1. Maxwell JC (1865) A dynamical theory of the electromagnetic field. Philos Trans R Soc Lond 155:459–512. doi:10.1098/rstl.1865.0008
2. Weiss P (1906) La variation du ferromagnetisme du temperature. Comptes Rendus 143:1136–1149
3. Stoner EC, Wohlfarth EP (1948) A mechanism of magnetic hysteresis in heterogeneous alloys. philosophical transactions of the royal society of London. Ser A Math Phys Sci 240 (826):599–642. doi:10.1098/rsta.1948.0007
4. Gerlach W, Stern O (1922) Das magnetische Moment des Silberatoms. Zeitschrift für Physik 9(1):353–355. doi:10.1007/bf01326984
5. Uhlenbeck GE, Goudsmit S (1926) Spinning electrons and the structure of spectra. Nature 117:264–265. doi:10.1038/117264a0
6. Heisenberg W, Pauli W (1929) Zur Quantendynamik der Wellenfelder. Zeitschrift für Physik 56(1):1–61. doi:10.1007/bf01340129
7. Skomski R (2003) Nanomagnetics. J Phys Condens Matter 15(20):R841. doi:10.1088/0953-8984/15/20/202
8. Bader SD (2006) Opportunities in nanomagnetism. Rev Mod Phys 78(1):1–15. doi:10.1103/RevModPhys.78.1
9. Žutić I, Fabian J, Das Sarma S (2004) Spintronics: fundamentals and applications. Rev Mod Phys 76(2):323–410. doi:10.1103/RevModPhys.76.323
10. Landau L, Lifshitz E (1935) On the theory of the dispersion of magnetic permeability in ferromagnetic bodies. Phys Z. Sowjetunion 8(153):101–114
11. Gilbert TL (2004) A phenomenological theory of damping in ferromagnetic materials. IEEE Trans Magn 40(6):3443–3449. doi:10.1109/tmag.2004.836740
12. Griffiths JHE (1946) Anomalous high-frequency resistance of ferromagnetic metals. Nature 158:670–671. doi:10.1038/158670a0
13. Kittel C (1948) On the theory of ferromagnetic resonance absorption. Phys Rev 73(2):155–161. doi:10.1103/PhysRev.73.155
14. Kirilyuk A, Kimel AV, Rasing T (2010) Ultrafast optical manipulation of magnetic order. Rev Mod Phys 82(3):2731–2784. doi: 10.1103/RevModPhys.82.2731
15. Beaurepaire E, Merle JC, Daunois A, Bigot JY (1996) Ultrafast spin dynamics in ferromagnetic nickel. Phys Rev Lett 76(22):4250–4253. doi:10.1103/PhysRevLett.76.4250
16. Koopmans B, Malinowski G, Dalla Longa F, Steiauf D, Faehnle M, Roth T, Cinchetti M, Aeschlimann M (2010) Explaining the paradoxical diversity of ultrafast laser-induced demagnetization. Nat Mater 9(3):259–265. doi:10.1038/nmat2593
17. Malinowski G, Longa FD, Rietjens JHH, Paluskar PV, Huijink R, Swagten HJM, Koopmans B (2008) Control of speed and efficiency of ultrafast demagnetization by direct transfer of spin angular momentum. Nat Phys 4(11):855–858. doi:10.1038/nphys1092
18. Bigot J-Y, Vomir M (2013) Ultrafast magnetization dynamics of nanostructures. Ann Phys 525(1–2):2–30. doi:10.1002/andp.201200199
19. Barman A, Haldar A (2014) Time-domain study of magnetization dynamics in magnetic thin films and micro-and nanostructures. In: Camley RE, Stamps R (eds) Solid State Phys, vol 65. Academic Press, Elsevier Inc., Burlington, pp 1–108. doi:http://dx.doi.org/10.1016/B978-0-12-800175-2.00001-7
20. Tserkovnyak Y, Brataas A, Bauer GEW, Halperin BI (2005) Nonlocal magnetization dynamics in ferromagnetic heterostructures. Rev Mod Phys 77(4):1375–1421. doi:10.1103/RevModPhys.77.1375
21. Demokritov SO, Hillebrands B, Slavin AN (2001) Brillouin light scattering studies of confined spin waves: linear and nonlinear confinement. Phys Rep 348(6):441–489. doi:10.1016/S0370-1573(00)00116-2

22. Barman A, Wang S, Maas JD, Hawkins AR, Kwon S, Liddle A, Bokor J, Schmidt H (2006) Magneto-optical observation of picosecond dynamics of single nanomagnets. Nano Lett 6 (12):2939–2944. doi:10.1021/nl0623457

23. Keatley PS, Kruglyak VV, Neudert A, Galaktionov EA, Hicken RJ, Childress JR, Katine JA (2008) Time-resolved Investigation of magnetization dynamics of arrays of nonellipsoidal nanomagnets with nonuniform ground states. Phys Rev B 78(21):214412. doi:10.1103/PhysRevB.78.214412

24. Rana B, Kumar D, Barman S, Pal S, Fukuma Y, Otani Y, Barman A (2011) Detection of picosecond magnetization dynamics of 50 nm magnetic dots down to the single dot regime. ACS Nano 5(12):9559–9565. doi:10.1021/nn202791g

25. Baibich MN, Broto JM, Fert A, Vandau FN, Petroff F, Eitenne P, Creuzet G, Friederich A, Chazelas J (1988) Giant magnetoresistance of (001)Fe/(001) Cr magnetic superlattices. Phys Rev Lett 61(21):2472–2475. doi:10.1103/PhysRevLett.61.2472

26. Binasch G, Grünberg P, Saurenbach F, Zinn W (1989) Enhanced magnetoresistance in layered magnetic-structures with antiferromagnetic interlayer exchange. Phys Rev B 39 (7):4828–4830. doi:10.1103/PhysRevB.39.4828

27. Berger L (1996) Emission of spin waves by a magnetic multilayer traversed by a current. Phys Rev B 54(13):9353–9358. doi:10.1103/PhysRevB.54.9353

28. Slonczewski JC (1996) Current-driven excitation of magnetic multilayers. J Magn Magn Matter 159(1–2):L1–L7. doi:10.1016/0304-8853(96)00062-5

29. Parkin SSP, Bhadra R, Roche KP (1991) Oscillatory magnetic exchange coupling through thin copper layers. Phys Rev Lett 66(16):2152–2155. doi:10.1103/PhysRevLett.66.2152

30. Parkin SSP (1993) Origin of enhanced magnetoresistance of magnetic multilayers— Spin-dependent scattering from magnetic interface states. Phys Rev Lett 71(10):1641–1644. doi:10.1103/PhysRevLett.71.1641

31. Tsoi M, Jansen AGM, Bass J, Chiang WC, Seck M, Tsoi V, Wyder P (1998) Excitation of a magnetic multilayer by an electric current. Phys Rev Lett 80(19):4281–4284. doi:10.1103/PhysRevLett.80.4281

32. Dyakonov MI, Perel VI (1971) Current-induced spin orientation of electrons in semiconductors. Phys Lett A 35(6):459–460. doi:http://dx.doi.org/10.1016/0375-9601(71)90196-4

33. Yu AB, Rashba EI (1984) Oscillatory effects and the magnetic susceptibility of carriers in inversion layers. J Phys C Solid State Phys 17(33):6039. doi:10.1088/0022-3719/17/33/015

34. Garello K, Miron IM, Avci CO, Freimuth F, Mokrousov Y, Bluegel S, Auffret S, Boulle O, Gaudin G, Gambardella P (2013) Symmetry and magnitude of spin-orbit torques in ferromagnetic heterostructures. Nat Nanotechnol 8(8):587–593. doi:10.1038/nnano.2013.145

35. Mangin S, Gottwald M, Lambert CH, Steil D, Uhlir V, Pang L, Hehn M, Alebrand S, Cinchetti M, Malinowski G, Fainman Y, Aeschlimann M, Fullerton EE (2014) Engineered materials for all-optical helicity-dependent magnetic switching. Nat Mater 13(3):287–293. doi:10.1038/nmat3864

36. Lambert CH, Mangin S, Varaprasad B, Takahashi YK, Hehn M, Cinchetti M, Malinowski G, Hono K, Fainman Y, Aeschlimann M, Fullerton EE (2014) All-optical control of ferromagnetic thin films and nanostructures. Science 345(6202):1337–1340. doi:10.1126/science.1253493

37. Apalkov D, Khvalkovskiy A, Watts S, Nikitin V, Tang X, Lottis D, Moon K, Luo X, Chen E, Ong A, Driskill-Smith A, Krounbi M (2013) Spin-transfer torque magnetic random access memory (STT-MRAM). J Emerg Technol Comput Syst 9(2):1–35. doi:10.1145/2463585.2463589

38. Kruglyak VV, Demokritov SO, Grundler D (2010) Magnonics. J Phys D Appl Phys 43 (26):260301. doi:10.1088/0022-3727/43/26/260301

39. Azzawi S, Ganguly A, Tokaç M, Rowan-Robinson RM, Sinha J, Hindmarch AT, Barman A, Atkinson D (2016) Evolution of damping in ferromagnetic/nonmagnetic thin film bilayers as a function of nonmagnetic layer thickness. Phys Rev B 93(5):054402. doi:10.1103/PhysRevB.93.054402

40. Ganguly A, Azzawi S, Saha S, King JA, Rowan-Robinson RM, Hindmarch AT, Sinha J, Atkinson D, Barman A (2015) Tunable magnetization dynamics in interfacially modified $Ni_{81}Fe_{19}$/Pt bilayer thin film microstructures. Sci Rep 5:17596. doi:10.1038/srep17596

41. King JA, Ganguly A, Burn DM, Pal S, Sallabank EA, Hase TPA, Hindmarch AT, Barman A, Atkinson D (2014) Local control of magnetic damping in ferromagnetic/non-magnetic bilayers by interfacial intermixing induced by focused ion-beam irradiation. Appl Phys Lett 104(24):242410. doi:10.1063/1.4883860

42. Ganguly A, Rowan-Robinson RM, Haldar A, Jaiswal S, Sinha J, Hindmarch AT, Atkinson DA, Barman A (2014) Time-domain detection of current controlled magnetization damping in Pt/$Ni_{81}Fe_{19}$ bilayer and determination of Pt spin Hall angle. Appl Phys Lett 105 (11):112409. doi:10.1063/1.4896277

43. Mondal S, Choudhury S, Jha N, Ganguly A, Sinha J, Barman A (2017) All-optical detection of spin Hall angle in W/CoFeB/SiO_2 heterostructures by varying thickness of the tungsten layer. Phys Rev B 96(5):054414. doi:10.1103/PhysRevB.96.054414

Chapter 2
Spin Dynamics

As briefly mentioned in the introduction of Chap. 1, the spin dynamics has been a topic of intense research to address several intriguing fundamental physics issues. In this chapter, we describe various time scales involved with spin dynamics in detail. Earlier, external stimuli such as electrical excitation and optical excitation have been used to trigger the spin dynamics in ferromagnetic thin films, multilayers, and nanostructures [1–5]. Here, we introduce and summarize the theoretical and experimental studies pertaining to ultrafast optical manipulation of spins in ferromagnetic thin films and nanostructures. Theoretically, numerous interesting concepts are involved in understanding the spin dynamics, such as conservation of angular momentum and the associated coherent spin–photon interaction, role of magnetic anisotropies, many body exchange interaction, and relationship between fluctuation and dissipation. We present an overview of the different experimental and theoretical approaches and try to provide a comprehensive picture within which the effects of light on the net magnetization, magnetic anisotropy in the case of ferromagnetic thin films, and confined magnetic structures may be understood. It is also important to mention that understanding these mechanisms has a close relation to technological applications, in particular magnetic storage technology [3, 6]. The magnetization dynamics processes are potentially useful for writing and retrieving information in magnetic storage media in the fastest possible ways.

2.1 Phenomenological Description

As discussed in Chap. 1, when an ultrafast laser pulse excites a ferromagnetic material, the aftereffect of the interaction between the laser pulse and the ferromagnetic material becomes interesting. The technique of pump–probe measurements revolutionized experimental research in the spin-dynamics area. The advent of pulsed, femtosecond lasers in the late 1980s and early 1990s made possible to observe the phenomena where the effects are not solely given by the length of the

© Springer International Publishing AG 2018   13
A. Barman and J. Sinha, *Spin Dynamics and Damping in Ferromagnetic Thin Films and Nanostructures*, https://doi.org/10.1007/978-3-319-66296-1_2

Fig. 2.1 Transient remanent longitudinal MOKE signal of a Ni(20 nm)/MgF$_2$(100 nm) film for 7 mJ cm^{-2} pump fluence. The signal is normalized to the signal measured in the absence of pump beam. The line is a guide to the eye. *Reprinted with permission from Ref. [10]. Copyright 1996 by the American Physical Society*

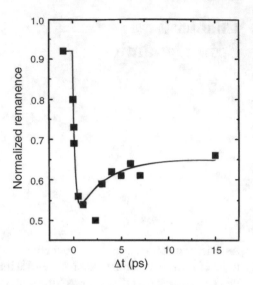

pump and probe pulses. Experiments pioneering the laser excitation of ferromagnetic metals were conducted by Vaterlaus et al. on Gd [7, 8]. However, they worked with pump and probe pulses with pulse width of 10 ns and 60 ps, respectively. In 1991, Freeman et al. reported the picosecond TR-MOKE measurement of magnetization dynamics of magnetic thin film [9]. The first observation of magnetization dynamics on the sub-picosecond timescale was made by Beaurepaire et al. in 1996 [10] as reproduced in Fig. 2.1. Experimentally, it was found that a nickel thin film can be demagnetized in a sub-picosecond time scale after excitation with a sub-100 fs laser pulse. During the fs laser excitation, the photons are absorbed instantaneously via certain electronic states that have a direct influence on magnetic parameters, such as, the magnetocrystalline anisotropy. One of the main goals in the field of laser-induced spin dynamics is to develop and understand a microscopic model which can successfully describe ultrafast demagnetization which till date is the fastest event following laser pulse excitation of ferromagnetic thin films. We discuss below phenomenological models which can, in general, describe the experimental observations considering the energy flux and the transfer of the angular momentum between three reservoirs: electrons, spins, and the lattice. These models do not include certain microscopic mechanisms which give rise to ultrafast demagnetization, and hence, still this is one of the ongoing research topics in this field.

(a) **Model Based on Rate Equations: Three-Temperature Model**

The three-temperature model (3TM) is an extension of the two-temperature model (2TM) which is used for describing the experimental investigation of picosecond laser pulse excitation in normal metals. The 2TM was first developed for characterizing normal metals when it is subjected to laser-induced carrier dynamics [11]. Briefly, in 2TM, the electron and lattice are considered as heat reservoirs, and it is assumed that these are coupled to exchange energy between them. In order to explain the magnetic system, Vaterlaus et al. [7] used the rate equation formulation by introducing the spin as one of the heat reservoirs. Within this model, the electron temperature T_e and lattice temperatures T_l characterize the electron and phonon distributions, whereas spin temperature T_s corresponds to spin distributions. In 3TM, it is assumed that the system consists of three thermalized reservoirs for exchanging energy, namely the electron, lattice, and spin systems at temperatures T_e, T_l, and T_s, respectively (Fig. 2.2). After the pump laser excitation of the magnetic thin films, within the first tens of fs, the experimentally observed non-equilibrium electron distribution cannot be described by a Fermi–Dirac distribution and thus no electron temperature can be derived. The absorbed energy creates hot electrons within the system, and during this transient hot electron regime, spin-dependent electron scattering modifies the spin population. This model was later invoked by Beaurepaire et al. [10] to explain the experimental observation of ultrafast demagnetization as shown in Fig. 2.1. It is assumed that during this process, the spin dynamics associated with T_s is induced which is different from T_e to T_l. Finally, this results in ultrafast demagnetization of the ferromagnetic material. The temporal evolution of the system can be described by three coupled differential equations:

$$C_e(T_e)\frac{\mathrm{d}T_e}{\mathrm{d}t} = -g_{el}(T_e - T_l) - g_{es}(T_e - T_s) + P(t) \tag{2.1}$$

$$C_s(T_s)\frac{\mathrm{d}T_s}{\mathrm{d}t} = -g_{es}(T_s - T_e) - g_{sl}(T_s - T_l) \tag{2.2}$$

$$C_l(T_l)\frac{\mathrm{d}T_l}{\mathrm{d}t} = -g_{el}(T_l - T_e) - g_{sl}(T_l - T_s) \tag{2.3}$$

where C primarily refers to the specific heat, g refers to the coupling constant between the reservoirs, and the laser heating of the electron system is introduced using the term $P(t)$. More details of the symbols are listed below:

C_e Electronic specific heat of the material concerned
C_s Magnetic contribution to the specific heat
C_l Lattice contribution to the specific heat
g_{el} Electron–lattice interaction constant
g_{sl} Spin–lattice interaction constant
g_{es} Electron–spin interaction constant
$P(t)$ Laser source term

The 3TM intuitively describes the energy equilibration processes during ultrafast demagnetization and the recovery of the magnetization back to equilibrium conditions. Using analytical solutions of the rate Eqs. (2.1) to (2.3), experimental demagnetization data has been fit by various groups. In order to simulate the energy equilibration processes between electrons, spins, and the lattice, input from experimental data, for example, the electron–lattice thermalization time from

Fig. 2.2 Schematic
representing the electron,
lattice, and spin reservoirs

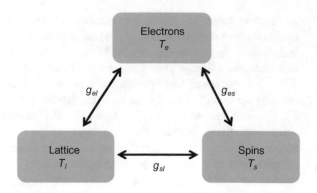

reflectivity measurements is used. Readers may refer to interesting literatures for
further details as we intend to provide an intuitive picture of the involved mechanisms without going into deep mathematical calculations [1]. It is important to
mention here that study of ultrafast demagnetization has been performed in several
materials where more than one magnetic constituents are present, for example,
$Ni_{80}Fe_{20}$, CoPt, and Heusler alloys [12, 13]. In 2012, Mentink et al. developed a
new model which explicitly includes the dynamics of multiple sublattices in a
magnetic material of relativistic and exchange origin [14].

Though the 3TM can satisfactorily describe some experimental observations, but
it is still a subject of intense debate due to some of its limitations. We further
discuss other models that have been used to explain ultrafast demagnetization of
ferromagnetic thin films which explicitly account for the transfer of spin angular
momentum during ultrafast demagnetization by proposing photons and phonons as
a reservoir for angular momentum.

(b) **Model Based on Electron–Phonon Spin-Flip Scattering**

In 2005, Koopmans et al. proposed that during ultrafast demagnetization,
phonon-mediated spin-flip scattering is responsible for the spin angular momentum
transfer to the lattice [15]. It was presumed that this spin-flip scattering is of the
Elliot–Yafet (EY) [16, 17] type found in paramagnetic metals. It is further assumed
that each electron–phonon scattering event can lead to a spin flip with a probability
α_{EY}, which is material dependent. By extending the phenomenological 3TM with
EY-like spin-flip channel, Koopmans named it as a microscopic three-temperature
model (M3TM). Within this model, using ab initio density functional theory, the
underlying spin-flip probability was calculated. A major limitation of M3TM was
discussed by Carva et al. in 2011 [18], where it was shown the demagnetization rate
in thermalized electron distributions as assumed under M3TM was significantly
small. In another study in 2011, Essert et al. [19] surmised that the dynamical
changes of the band structure play important role in the ultrafast demagnetization
instead of electron–phonon spin-flip scattering as presumed in M3TM. Prior to
these findings, in 2008, Carpene et al. [20] had proposed a spin-flip mechanism
similar to the M3TM, but with magnons acting as the reservoir for spin angular

momentum, whereas, in 2009, Kraub et al. [21] had suggested a microscopic demagnetization mechanism based on electron–electron Coulomb scattering. Using these models, ultrafast demagnetization results of Ni and Co were reproduced by considering interband scattering processes which leads to a redistribution of electrons from majority to minority bands after the optical excitation of spin-polarized electrons.

(c) **Model Based on Interaction Between Photons and Spins**

Zhang and Hübner in 2000 [22] proposed a model for ultrafast demagnetization as a cooperative effect in the presence of both an external laser field and the internal spin–orbit coupling. It was concluded that in the absence of spin–orbit coupling, the laser field alone cannot change the magnetic moment on fs timescale; as well as without the laser field, the spin–orbit coupling can cause only few percent change in the magnetic moment. A major drawback of this model is the assumption that the pump laser photons serve as the source for the angular momentum needed to flip the spins in the ferromagnet. Under this condition, the mechanism of demagnetization should vary depending on the polarization of the pump laser beam if it is linearly or circularly polarized. Within a few years of the prediction by Zhang and Hübner, in 2007, Longa et al. [23] found that the magnetization dynamics remain unchanged by introducing any polarization-dependent changes in time-resolved magneto-optical Kerr effect (TR-MOKE) measurements. Furthermore, the model of Zhang and Hübner suggested that the ultrafast demagnetization has quasi-instantaneous nature and the observed time constant is given by the pump pulse length which remained controversial to be verified in the experiments. In 2009, Bigot et al. [24] proposed a similar microscopic mechanism in which a laser-field-induced time-dependent modification of the spin–orbit interaction was taken into account. Consideration of such effects leads to a coherent interaction between the pump photons and the spins. During this process, the transfer of angular momentum is necessary for the spin flips to occur.

(d) **Model Based on Superdiffusive Spin Transport**

A relatively advanced model to describe the origin of ultrafast demagnetization has been proposed by Battiato et al. in 2010 which takes into account the 'superdiffusive spin transport' [25]. Within this model, the reduction of magnetization is explained by spin-dependent transport of charge carriers out of a ferromagnetic layer instead of spin flips. It is assumed that due to pump laser irradiation of a ferromagnetic layer, electrons are excited from quasi-localized d-bands to more mobile sp-bands, and this process is spin conserving. Subsequently, electrons are transported superdiffusively out of the excited sample volume into the substrate. This phenomenon is referred as 'superdiffusive' due to the consideration of typical carrier transport which has ballistic character initially but converges to diffusive transport over longer time scale. Thus, for all timescales, either the ballistic or diffusive approximation fails to describe the carrier transport. It has been conjectured that the superdiffusive transport gives rise to ultrafast demagnetization possibly due to the following mechanisms: (1) laser-excited electrons in *sp*-bands have high velocities, and (2) the lifetimes of the excited spin majority and minority electrons are different. As a result of the second point, the excited majority carriers in 3*d* ferromagnet are more mobile than the minority carriers which may lead to a deficit of majority carriers in the magnetic film and a transfer of magnetization away from the surface. For further details, readers may refer to detailed description in other relevant references [26, 27]. The next section describes the magnetization dynamics at various time-scales.

2.1.1 Magnetization Dynamics at Various Time Scales

A glimpse into the involved time scales in magnetism shows a very broad range varying from femtosecond to microsecond scale [1]. In the previous section, we described the ultrafast demagnetization which occurs during the first few hundreds of fs after laser irradiation of ferromagnetic thin films. It has been well established that the magnetization dynamics takes place on time scales that are usually longer than the time necessary in a ferromagnetic thin film to recover equilibrium between the carriers and the spin reservoir after the laser irradiation. The quantities such as saturation magnetization and effective anisotropy may be considered as time invariant over long time beyond fs regime. The precession of magnetization occurs within 10–100 ps and gets damped in sub-ns to tens of ns time [28]. Within similar time scales, two more phenomena, namely reversal of spins as used in magnetic recording (few ps—few hundreds of ps) and vortex core switching (few tens of ps —several ns), are known to happen. The slowest dynamics is the domain wall motion. The typical time scale of this process is few ns to few µs. One of the motivations here is to understand the magnetization dynamics beyond the ultrafast demagnetization regime in ferromagnetic thin films and nanostructures with a particular focus on possibility of investigating magnetic damping. Earlier experimental studies showed that an ultrafast magnetic response can be used to trigger coherent precession of magnetization. In ferromagnetic materials, this study is considered as a first step toward an ultrafast and coherent optical control of magnetization which is significantly important for magnetic recording industry. Another interesting experimental study reported that the laser-induced demagnetization could be advantageous to trigger the magnetization precession in a ferromagnetic/antiferromagnetic exchange coupled system [29]. It was shown that the exchange bias between magnetic layers is modified by the ultrafast demagnetization to the point that the spin angular momentum is affected, leading to the precession of the magnetization on a time scale of a few hundreds of ps. The resulting process is coherent and decays within the characteristic dephasing time of the electronic levels. This discussed mechanism has been intensely used after this finding to optically induce the ferromagnetic resonance in the absence of any external radio frequency magnetic field.

2.1.2 Optically Induced Ultrafast Spin Dynamics

The unique feature of optically induced ultrafast spin dynamics is it allows accessing the dynamics over fs to ns time scale. Intense research on experimental and theoretical fronts was pursued after the seminal work of observation of ultrafast demagnetization by Beaurepaire et al. in 1996 [10]. Koopmans et al., in 2000, debated over the origin of observed magneto-optical signal obtained during magnetization dynamics measurement in Ni [30]. The main concern of these authors was whether the experimental signal observed has purely magnetic origin or it also contains optical contributions. In their study, the authors explicitly measured the time-resolved Kerr ellipticity and rotation, as well as its temperature and magnetic field dependence in epitaxially grown (111)- and (001)-oriented Cu/Ni/Cu wedge-shaped thin films with Ni thickness varying from 0 to 15 nm. Their main finding was that in the first hundreds of fs, the response is dominated by state-filling effects, whereas the actual demagnetization occurs in approximately 0.5–1 ps, and in the sub-ns time scale regime, the spins precess in the anisotropy field. Within two years of this study, in 2002, van Kampen et al. [28] showed a novel all-optical method of excitation and detection of spin waves which is considered to be a major breakthrough in generalizing the laser-induced magnetization dynamics in ferromagnetic thin films. Despite of the fact that the above-mentioned phenomenon is described in the existing literatures, for the sake of completeness of the discussion, here, we emphasize the damping aspect of the experimental result (cf. Fig. 2.3). In this study, a polycrystalline Ni thin film deposited on silicon substrate with a canted equilibrium orientation of magnetization in an external applied field was investigated using a pump–probe-based TR-MOKE magnetometer. The schematic of the experimental setup is shown in Fig. 2.3a with an arrangement of focusing the pump and probe laser beams on the sample along with the provision of external magnetic field application. When the pump pulse hits the Ni film, a sharp decrease in the perpendicular component of magnetization M_z is observed. Primarily, the change in magnitude of the (temperature-dependent) magnetization is responsible for such decrease. It is followed by a subsequent recovery of M_z on a time scale of a few ps which is due to rapid heat diffusion into the substrate. Remarkably, over few hundreds of ps post-thermal equilibrium, a secondary response appears in the TR-MOKE signal as a persistent oscillation as shown in Fig. 2.3b. A delicate balance between the external field and the net anisotropy field of the film determines the canting angle. The transient heating of the Ni film due to pump pulse also results in the change in anisotropy of the film. This causes a change of the equilibrium orientation of M from θ_c to $\theta_c{'}$ as indicated in Fig. 2.3c (IIa), triggering an initial precession of the magnetization around its new equilibrium orientation as schematically represented in (IIb). For the case of metallic films, the original equilibrium angle is restored after around 10 ps of removal of excess heat by heat diffusion into the thin film. It is interesting to note here that up to this point, the magnetization is still not in equilibrium due to its initial displacement. Thus, the magnetization continues to precess for few hundreds of ps as shown in scheme (III).

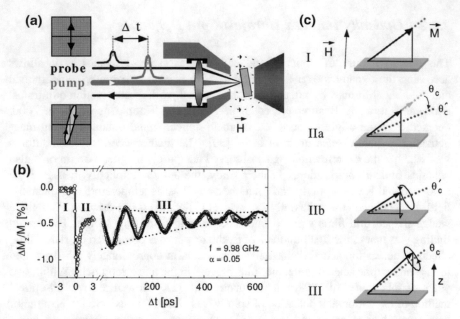

Fig. 2.3 Schematic pump-probe setup. **a** The magnetization is measured by the polarization state of the reflected probe pulse. **b** Typical measurement on a 7 nm Ni film (open circles: data; thick line: fit) displaying the perpendicular component of the magnetization, M_Z, as a function of delay time Δt. The different stages are indicated by numbers. **c** The stages of the excitation process: (I) $\Delta t < 0$, the magnetization $\vec{M}$ points in equilibrium direction (dotted line), (IIa) $\Delta t = 0$, the magnitude of $\vec{M}$ and the anisotropy change due to heating, thereby altering the equilibrium orientation, (IIb) $0 < \Delta t < 10$ ps, $\vec{M}$ starts to precess around its new equilibrium, (III) $\Delta t > 10$ ps, heat has diffused away, the magnitude of $\vec{M}$ and anisotropy are restored, but the precession continues because of the initial displacement of $\vec{M}$. *Reprinted with permission from Ref.* [28]. *Copyright 2002 by the American Physical Society*

However, in a real system, energetically it is not favored for the magnetization to precess for infinitely long duration. Importantly, the effective damping (α_{eff}) governs the magnetization dynamics till the time the magnetization relaxation occurs. Furthermore, in this study, it was found that the laser-induced spin precession in single magnetic layer has equivalence with the frequency as observed in a 'conventional ferromagnetic resonance (FMR)' experiment, and the damping constant of the Ni film estimated using TR-MOKE and FMR technique is consistent. This led to the establishment of a sensitive optical excitation and detection tool which may be used to study coherent magnetization dynamics including important aspects of damping phenomena in ferromagnetic thin films. An important question which kept puzzling the research community working in this field was if optical excitation and detection-based technique can be implemented for investigating the magnetization dynamics, in particular, for estimating the effective damping coefficient 'α_{eff}', in magnetic nanostructures.

In 2006, Barman et al. [4] developed the technique of Cavity Enhanced-TR-MOKE (more details of this technique will be discussed in Chap. 4 of this book) to investigate ultrafast magnetization dynamics in ferromagnetic nano-elements. Later in 2007 [31], the authors reported a size-dependent effective damping (α_{eff}) of nickel nanomagnets (cylindrical nickel dots with constant height of 150 nm and with varying diameter D from 5 μm down to 125 nm) extracted directly from the time-resolved Kerr rotation. The dots were coated with 70-nm-thick SiN layer to have a fivefold increase in the Kerr rotation due to cavity enhancement. Traces of the TR-MOKE data as observed in their study are reproduced in Fig. 2.4a. High-pass fast Fourier transform filtering of the experimental time-resolved data was performed to eliminate any low-frequency background. The filtered time-resolved data was shown to possess a single uniform precession frequency, and these were fitted through a least square fitting routine with a damped sine function of the form

$$M(t) = M(0)e^{-t/\tau}\sin(\omega t - \phi) \tag{2.4}$$

where τ is the decay time of the precession defined as $\tau = 1/\omega\alpha_{eff}$, ω is the angular frequency of the uniform precession mode given by $\omega = \gamma H_{eff}$, ϕ is the initial phase of oscillation, γ is the gyromagnetic ratio, and H_{eff} is the effective magnetic field. Briefly, in this study, large qualitative and quantitative differences in between the microscale and nanoscale magnetic nanodots were observed (Fig. 2.4b). It was found that for nanomagnet of diameter 2 μm, α_{eff} reduces sharply. Furthermore, a slow decrease in α_{eff} down to 500 nm was observed where it attains a value 0.04, comparable to the reported damping coefficient 0.05 of continuous Ni thin film measured by all-optical method. In Fig. 2.4c, a representative time-resolved data from a magnetic dot of 400 nm diameter at varying external bias field is shown. It was inferred from these data that the α_{eff} almost remains invariant with respect to external bias field. Overall, they reported that for the magnetic elements with $D > 2$ μm a strong bias field dependence of α_{eff} was observed, while for those with $D < 1$ μm, no bias field dependence was observed. For magnetic elements with intermediate D, weak bias field dependence was observed. The above observations established that the all-optical excitation and detection are well suited for investigating the magnetization dynamics, including damping behavior in magnetic nanostructures. It is worth mentioning here that the research on laser-induced magnetization dynamics, in particular, the ultrafast demagnetization, different relaxation mechanisms, and investigation of precession frequency and effective damping behavior have grown significantly over the last decade and to cover all details of those is beyond the scope of this book.

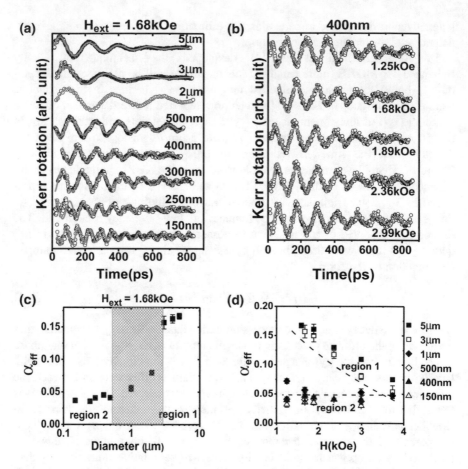

Fig. 2.4 **a** Experimental (*open circles*) and fitted (*gray lines*) time-resolved data from magnetic dots of varying diameter at an external bias field = 1.68 kOe. **b** The extracted effective damping coefficient α_{eff} as a function of magnet diameter. The hatched rectangle shows the transition region from a high to low α_{eff}. **c** Experimental (*open circles*) and simulated (*gray lines*) time-resolved data from a magnetic dot of 400 nm diameter at varying external bias fields. **d** The extracted α_{eff} for magnets with varying diameter as a function of the external bias field. A high-pass FFT filtering was applied to the experimental time-resolved data for fitting with a single damped sine function. *Reprinted with permission from Ref.* [31]. *Copyright 2007 by the American Institute of Physics*

2.1.3 Landau–Lifshitz–Gilbert (LLG) Equation

Following quantum mechanics, equation of motion for the dynamics of single spin can be derived, and it is known as Landau–Lifshitz (LL) [32] equation mentioned as below

$$\frac{\mathrm{d}}{\mathrm{d}t} \langle S \rangle = \frac{g\mu_B}{\hbar} \langle S \times B \rangle \tag{2.5}$$

where S denotes spin and B the magnetic field, and the multiplicative term corresponds to the gyromagnetic ratio $\gamma = g\mu_B/\hbar$. In the macroscopic model, the magnetization vector is defined by uniform distribution of spin in the sample, and it is expressed as below:

$$M = -\frac{g\mu_B}{\hbar}\langle S\rangle \qquad (2.6)$$

Therefore, equation of motion of magnetization in presence of external magnetic field is given by

$$\frac{dM}{dt} = -\frac{g\mu_B}{\hbar}\langle M \times H\rangle \qquad (2.7)$$

Equation (2.7) is known as Landau–Lifshitz (LL) equation, and this is generalized by using H_{eff} in place of H. The Landau–Lifshitz model considers that the total magnetic moment is related to the total angular momentum which experiences the net torque resulting in a precessional motion. It thus implies that the tip of the magnetization vector precesses around the effective magnetic field in a circular orbit as shown in Fig. 2.5a for infinite duration with angular frequency $\omega = \gamma H_{\text{eff}}$. However, in all practical situations, the precession amplitude of magnetization decreases with time, and the tip of the magnetization vector follows a spiral path as shown in Fig. 2.5b. Thus, in order to describe real magnetization dynamics, a damping term needs to be included in the LL equation.

For practical systems, Gilbert [33, 34] proposed a phenomenological term used to describe the damping of the magnetization precession. Depending on the time derivative of the magnetization, Gilbert modeled a 'viscous' damping by

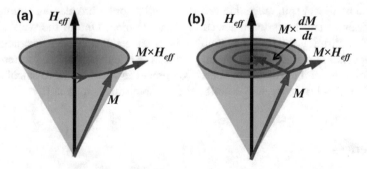

Fig. 2.5 Precession of magnetization about the applied bias field **a** without damping and **b** with damping

considering a phenomenological dissipation term. The precessional dynamics is described by Landau–Lifshitz–Gilbert (LLG) Eq. (2.8)

$$\frac{d\vec{M}}{dt} = -\gamma(\vec{M} \times \vec{H}_{\text{eff}}) + \frac{\alpha}{M}\left(\vec{M} \times \frac{d\vec{M}}{dt}\right) \tag{2.8}$$

where α the dimensionless Gilbert damping and γ the gyromagnetic ratio. The effective magnetic field

$$H_{\text{eff}} = H_a + H_{\text{dem}} + H_{\text{ext}} + H_{\text{exc}}$$

is composed of anisotropy fields (H_a), the demagnetizing field (H_{dem}), external applied magnetic field (H_{ext}), and exchange field (H_{exc}). Importantly, similar to the LL model, the magnitude of the magnetization is conserved in the LLG equation as well. To reiterate, the Gilbert damping term was specifically introduced to replicate the experimental observation in the best conceivable manner. The LLG equation has been used to study the magnetization dynamics in various magnetic systems. If the magnetic system is sufficiently small, then the magnetization may be assumed to remain uniform during the dynamics, and anisotropy field, demagnetizing field, and the applied external field contribute to the effective field. For larger samples and in the case of inhomogeneous dynamics, the magnetic moment becomes a function of spatial coordinates. The effective magnetic field in this case also acquires a contribution from the exchange interaction. In such situation, non-homogeneous elementary excitations of the magnetic medium may exist, first proposed by Felix Bloch in 1930 [35]. These excitations are called spin waves and involve many lattice sites. More details on these aspects can be found in existing literatures [36]. Commonly, it has been thought that intrinsic Gilbert damping had its origin in spin–orbit coupling because this mechanism does not conserve spin, but it has never been derived from a coherent framework. For ferromagnetic thin films and nanostructures, the non-local spin relaxation processes and disorder broadening couple to the spin dynamics and leads to enhanced Gilbert damping. One of the well accepted ways of confirming the form of damping is to understand the ferromagnetic resonance spectral linewidth. The damping constant is often reformulated in terms of a relaxation time, and the dominant relaxation processes are invoked to calculate it. In Chap. 3 of this book, we will describe the damping in more detail.

References

1. Kirilyuk A, Kimel AV, Rasing T (2010) Ultrafast optical manipulation of magnetic order. Rev Mod Phys 82(3):2731–2784. doi:10.1103/RevModPhys.82.2731
2. Tserkovnyak Y, Brataas A, Bauer GEW, Halperin BI (2005) Nonlocal magnetization dynamics in ferromagnetic heterostructures. Rev Mod Phys 77(4):1375–1421. doi:10.1103/RevModPhys.77.1375

3. Hoffmann A, Bader SD (2015) Opportunities at the Frontiers of spintronics. Phys Rev Appl 4 (4):047001. doi:10.1103/PhysRevApplied.4.047001
4. Barman A, Wang S, Maas JD, Hawkins AR, Kwon S, Liddle A, Bokor J, Schmidt H (2006) Magneto-optical observation of picosecond dynamics of single nanomagnets. Nano Lett 6 (12):2939–2944. doi:10.1021/nl0623457
5. Liu L, Pai C-F, Li Y, Tseng HW, Ralph DC, Buhrman RA (2012) Spin-torque switching with the giant spin Hall effect of tantalum. Science 336(6081):555–558. doi:10.1126/science.1218197
6. Bader SD (2006) Opportunities in nanomagnetism. Rev Mod Phys 78(1):1–15. doi:10.1103/RevModPhys.78.1
7. Vaterlaus A, Beutler T, Meier F (1991) Spin-lattice relaxation time of ferromagnetic gadolinium determined with time-resolved spin-polarized photoemission. Phys Rev Lett 67 (23):3314–3317. doi:10.1103/PhysRevLett.67.3314
8. Vaterlaus A, Beutler T, Guarisco D, Lutz M, Meier F (1992) Spin-lattice relaxation in ferromagnets studied by time-resolved spin-polarized photoemission. Phys Rev B 46 (9):5280–5286. doi:10.1103/PhysRevB.46.5280
9. Freeman MR, Ruf RR, Gambino RJ (1991) Picosecond pulsed magnetic fields for studies of ultrafast magnetic phenomena. IEEE Trans Magn 27(6):4840–4842. doi:10.1109/20.278964
10. Beaurepaire E, Merle JC, Daunois A, Bigot JY (1996) Ultrafast spin dynamics in ferromagnetic nickel. Phys Rev Lett 76(22):4250–4253. doi:10.1103/PhysRevLett.76.4250
11. Kaganov MI, Lifshitz IM, Tanatarov LV (1957) Relaxation between electrons and the crystalline lattice. J Exp Theor Phys 4(2):173
12. Guidoni L, Beaurepaire E, Bigot J-Y (2002) Magneto-optics in the ultrafast regime: thermalization of spin populations in ferromagnetic films. Phys Rev Lett 89(1):017401. doi:10.1103/PhysRevLett.89.017401
13. Mann A, Walowski J, Münzenberg M, Maat S, Carey MJ, Childress JR, Mewes C, Ebke D, Drewello V, Reiss G, Thomas A (2012) Insights into ultrafast demagnetization in pseudogap half-metals. Phys Rev X 2(4):041008. doi:10.1103/PhysRevX.2.041008
14. Mentink JH, Hellsvik J, Afanasiev DV, Ivanov BA, Kirilyuk A, Kimel AV, Eriksson O, Katsnelson MI, Rasing T (2012) Ultrafast spin dynamics in multisublattice magnets. Phys Rev Lett 108(5):057202. doi:10.1103/PhysRevLett.108.057202
15. Koopmans B, Ruigrok JJM, Longa FD, de Jonge WJM (2005) Unifying ultrafast magnetization dynamics. Phys Rev Lett 95(26):267207. doi:10.1103/PhysRevLett.95.267207
16. Elliott RJ (1954) Theory of the effect of spin-orbit coupling on magnetic resonance in some semiconductors. Phys Rev 96(2):266–279. doi:10.1103/PhysRev.96.266
17. Yafet Y (1963) Solid state phys vol. 14. Academic Press
18. Carva K, Battiato M, Oppeneer PM (2011) Ab initio investigation of the Elliott-Yafet electron-phonon mechanism in laser-induced ultrafast demagnetization. Phys Rev Lett 107 (20):207201. doi:10.1103/PhysRevLett.107.207201
19. Essert S, Schneider HC (2011) Electron-phonon scattering dynamics in ferromagnetic metals and their influence on ultrafast demagnetization processes. Phys Rev B 84(22):224405. doi:10.1103/PhysRevB.84.224405
20. Carpene E, Mancini E, Dallera C, Brenna M, Puppin E, De Silvestri S (2008) Dynamics of electron-magnon interaction and ultrafast demagnetization in thin iron films. Phys Rev B 78 (17):174422. doi:10.1103/PhysRevB.78.174422
21. Krauß M, Roth T, Alebrand S, Steil D, Cinchetti M, Aeschlimann M, Schneider HC (2009) Ultrafast demagnetization of ferromagnetic transition metals: the role of the Coulomb interaction. Phys Rev B 80(18):180407. doi:10.1103/PhysRevB.80.180407
22. Zhang GP, Hübner W (2000) Laser-induced ultrafast demagnetization in ferromagnetic metals. Phys Rev Lett 85(14):3025–3028. doi:10.1103/PhysRevLett.85.3025
23. Dalla Longa F, Kohlhepp JT, de Jonge WJM, Koopmans B (2007) Influence of photon angular momentum on ultrafast demagnetization in nickel. Phys Rev B 75(22). doi:10.1103/PhysRevB.75.224431

24. Bigot JY, Vomir M, Beaurepaire E (2009) Coherent ultrafast magnetism induced by femtosecond laser pulses. Nat Phys 5(7):515–520. doi:10.1038/nphys1285
25. Battiato M, Carva K, Oppeneer PM (2010) Superdiffusive spin transport as a mechanism of ultrafast demagnetization. Phys Rev Lett 105(2):027203. doi:10.1103/PhysRevLett.105.027203
26. Carva K, Battiato M, Oppeneer PM (2011) Is the controversy over femtosecond magneto-optics really solved? Nat Phys 7(9):665. doi:10.1038/nphys2067
27. Eschenlohr A, Battiato M, Maldonado R, Pontius N, Kachel T, Holldack K, Mitzner R, Fohlisch A, Oppeneer PM, Stamm C (2013) Ultrafast spin transport as key to femtosecond demagnetization. Nat Mater 12(4):332–336. doi:10.1038/nmat3546
28. van Kampen M, Jozsa C, Kohlhepp JT, LeClair P, Lagae L, de Jonge WJM, Koopmans B (2002) All-optical probe of coherent spin waves. Phys Rev Lett 88(22):227201. doi:10.1103/PhysRevLett.88.227201
29. Ju G, Nurmikko AV, Farrow RFC, Marks RF, Carey MJ, Gurney BA (1999) Ultrafast time resolved photoinduced magnetization rotation in a ferromagnetic/antiferromagnetic exchange coupled system. Phys Rev Lett 82(18):3705–3708. doi:10.1103/PhysRevLett.82.3705
30. Koopmans B, van Kampen M, Kohlhepp JT, de Jonge WJM (2000) Ultrafast magneto-optics in nickel: magnetism or optics? Phys Rev Lett 85(4):844–847. doi:10.1103/PhysRevLett.85.844
31. Barman A, Wang S, Maas J, Hawkins AR, Kwon S, Bokor J, Liddle A, Schmidt H (2007) Size dependent damping in picosecond dynamics of single nanomagnets. Appl Phys Lett 90:202504. doi:10.1063/1.2740588
32. Landau L, Lifshitz E (1935) On the theory of the dispersion of magnetic permeability in ferromagnetic bodies. Physikalische Zeitschrift der Sowjetunion 8(153):101–114
33. Gilbert TL (2004) A phenomenological theory of damping in ferromagnetic materials. IEEE Trans Magn 40(6):3443–3449. doi:10.1109/tmag.2004.836740
34. Gilbert TL (1955) A Lagrangian formulation of the gyromagnetic equation of the magnetic field. Phys Rev 100:1243
35. Bloch F (1930) Zur Theorie des Ferromagnetismus. Zeitschrift für Physik 61:206
36. Demokritov SO, Hillebrands B, Slavin AN (2001) Brillouin light scattering studies of confined spin waves: linear and nonlinear confinement. Phys Rep 348(6):441–489. doi:10.1016/S0370-1573(00)00116-2

Chapter 3
Magnetic Damping

In this chapter, we focus on various aspects of magnetic damping. As described in the previous chapters, in a ferromagnetic system, the spin dynamics is described by using Landau–Lifshitz–Gilbert (LLG) equation in which a phenomenological parameter α defines a magnetization relaxation. In 1956, Gilbert proposed that the LLG equation may be phenomenologically derived for an electron gas in a strong molecular field assuming that spin states relax to an instantaneous equilibrium at a rate (τ_s) expressed as $(\lambda \approx 1/\beta\tau_s)$ where β corresponds to the classical molecular field coefficient [1, 2]. In general, it is understood that the magnetic damping occurs because of the coupling of the magnetic modes (primarily the electron spins) to the non-magnetic modes (the electron orbits and lattice vibrations) in the magnetic system [3, 4]. Due to this, there is a back-and-forth transfer of energy and the flow of energy predominantly occurs from the magnetic modes to the non-magnetic modes as the magnetic modes are typically excited to a higher temperature than the other modes in the ferromagnetic system. Understanding various coupling mechanisms and quantifying their contributions to damping has been a perennial problem for decades especially after Landau and Lifshitz [5] published their phenomenological equation of motion in 1935. In this chapter, we provide a brief overview of some of damping mechanisms that have been studied with accepting the fact that this field is far too rich and vast to cover each and every detail here.

In Chap. 1, some of the experimental techniques used to probe magnetization dynamics and damping have been mentioned. To briefly recapitulate, historically, the magnetic damping has been probed in the frequency domain primarily using the ferromagnetic resonance (FMR) technique [6, 7]. The magnetic damping controls the dissipation rate associated with the small amplitude magnetization motion as probed in FMR. By measuring the resonance linewidth in the FMR experiment (half width at half maximum), $\Delta H(\omega)$ as a function of microwave frequency ω, the Gilbert damping (α) of a ferromagnetic material can be expressed as:

© Springer International Publishing AG 2018
A. Barman and J. Sinha, *Spin Dynamics and Damping in Ferromagnetic Thin Films and Nanostructures*, https://doi.org/10.1007/978-3-319-66296-1_3

$$\Delta H(\omega) = \alpha \frac{2\pi\omega}{\gamma} \tag{3.1}$$

where γ is the gyromagnetic ratio. There has been intense effort to understand the linewidth broadening mechanism in FMR which is the measure of magnetization damping [7–9]. A common consensus is that the linewidth broadening in FMR can be decomposed into intrinsic and extrinsic mechanisms [4]. The basis of this categorization has been, whether a mechanism responsible for the damping is inevitable or avoidable. Briefly, the intrinsic damping (inevitable) originates primarily from interactions between the spins and the electron orbits, which is quite fundamental in nature and includes generation of eddy currents and spin–orbit coupling. In addition to this, intrinsic contribution to damping occurs due to coupling of the uniform precession with the radiation field and through direct magnon–phonon scattering. On the other hand, extrinsic contribution to damping can be avoided in a perfect sample. It mainly arises due to sample inhomogeneities, which lead to linewidth broadening in the FMR experiments [10]. Primarily, the extrinsic contribution to damping results in producing a distribution of local resonance fields across the sample. Below, we discuss the basics of intrinsic and extrinsic damping in more detail without bringing in the rigorous mathematical calculations.

3.1   Intrinsic Contribution

Intrinsic effects cause damping even in ferromagnetic materials which have perfect crystal structure. In a broad sense, in the perfect systems there are certain interactions between the magnetization and the other degrees of freedom. Thus, the coupling between these allows energy to leak from the magnetic system to the other systems, e.g., lattice. It is understood that the spins couple to magnetic fields in similar way as the electron eddy currents and atomic orbits are coupled. Briefly, the magnetic relaxation processes that involve the electron scattering with phonons and thermally excited magnons are considered as the source of intrinsic contribution to the damping. From the perspective of experimental determination of intrinsic damping, it is quite challenging to provide precise method as it often leads to intense debate. A somewhat common understanding up to few decades back was, under well-defined thermodynamic condition, the smallest measured FMR linewidth correspond to intrinsic damping of the system [4]. With the development of time-resolved magneto-optical Kerr effect in recent time, it has been argued as a sensitive technique to provide an unambiguous estimate of the intrinsic damping [11, 12]. Below we discuss the mechanisms, which are responsible for intrinsic damping.

3.1.1 Spin–Orbit Coupling

It is reasonable to consider within approximations that the electrons in an atom have well-defined orbital and spin angular momentum, and these are defined as l_z and σ_z. An important consequence of spin–orbit interaction is, instead of l_z and σ_z remaining as good quantum number individually, $j_z = l_z + \sigma_z$ becomes good azimuthal quantum number. Due to this, the spin and orbital moments can exchange momentum back-and-forth during the precession of magnetization. Therefore, j precesses with constant j_z; however, the l_z and σ_z become misaligned and exert a torque on each other. This further results in precession of l_z and σ_z about j such that the corresponding z components no longer remain constant during magnetization precession. In the ferromagnetic resonance measurement, the spins get to an excited state; however, the orbits remain in the ground state. This results in pumping of energy from spin degrees of freedom into the orbital degree of freedom as j precesses. Interestingly, this process is frequently interrupted by electron–lattice scattering events. By removing energy from the electron and leaving the orbit in a lower energy configuration, these scattering events alter the orbital moment; however, the spin moment remains unperturbed. Overall, damping through the spin–orbit interaction can be considered as a two-step process in which the first step involves excitation of the orbital moments to a high-energy configuration by the spins, and subsequently, this energy is scattered to the lattice [3, 13].

In order to explain the intrinsic damping Kambersky, in 1976 [3], proposed a spin–orbit coupling (SOC) torque-correlation model by taking into account the contributions of intraband and interband transitions. It was found that these transitions play a dominant role in determining the intrinsic damping. For SOC strength ξ, at low and high temperatures the intrinsic damping was found to be proportional to ξ^3 and ξ^2, respectively. Further details of theoretical description of spin–orbit interaction on Gilbert damping is discussed in Sect. 3.3. In general, the intrinsic damping also gets affected by electron scattering time and density of states at Fermi surface which varies in various materials. Due to the difficulty in isolating the effects other than the SOC on intrinsic damping, it remained challenging for few decades to experimentally verify the theoretical prediction of Kambersky et al. In 2013, He et al. [14] in an interesting study reported the experimental verification of scaling of intrinsic damping with ξ^2 in $FePd_{1-x}Pt_x$ ternary alloy thin films by employing time-resolved magneto-optical Kerr effect. In this study, only ξ was modulated by changing Pd/Pt ratio as these heavy metals have large SOC while the other parameters remain fixed.

3.1.2 Phonon-Drag Mechanism

The direct phonon–magnon scattering is another possible relaxation mechanism. In 1998, Shul [15] proposed the mechanism of magnon relaxation by phonon drag, and this proposal is valid for small geometries where magnetization and lattice strain are homogeneous. According to the calculations of Shul [15], using LLG equation and the lattice strain equations, within the asymptotic limit, the Gilbert phonon damping is defined as

$$\alpha_{\text{phonon}} = \frac{2n\gamma}{M_s}\left(\frac{B_2(1+v)}{E}\right)^2 \tag{3.2}$$

where n is the phonon viscosity, B_2 is the magnetoelastic shear constant, E the Young's modulus and v is the Poisson ratio. In the above expression, except for phonon viscosity n, other parameters can be obtained following existing literatures. In 1982, Heinrich et al. [16] experimentally determined in microwave transmission experiments the phonon viscosity parameters for Ni crystals in microwave frequency range. In this interesting experiment, a fast transverse elastic shear wave was generated by magnetoelastic coupling inside the skin depth of a thick Ni (001) crystal at 9.5 GHz. They observed the reradiated microwave power on the opposite side of Ni slab due to the fact that the transverse elastic shear wave could propagate across the slab. This effect was termed as phonon-assisted microwave transmission. In this experiment, the elastic wave relaxation time was estimated to be 6.6×10^{-10} s at 9.5 GHz following the fit to the experimental data using LLG equation and elastic wave equations of motion including magnetoelastic coupling. The phonon viscosity is given by $n = c_{44}/\tau_{ph}\omega^2$, where c_{44} refers to as elastic modulus and for Ni, $n \approx 3.4$ in CGS unit. Thus, using Eq. 3.2, the phonon Gilbert damping is estimated as ~ 0.001 which is nearly 30 times smaller than the intrinsic Gilbert damping parameter for Ni. The phonon Gilbert damping is expected to be larger in case where magnetostrictive effect is large for the material.

3.1.3 Eddy-Current Mechanism

Eddy currents in a ferromagnetic material get induced by precession of magnetization. This mechanism is known as screening of the electromagnetic microwave field by the conduction electrons, which leads to another contribution to the magnetization relaxation, and dissipation in this process is proportional to the conductivity of the sample. Eddy currents become important when the film thickness is larger than or comparable to the skin depth (depth below the surface of the conductor over which the current density decays to $1/e$ value that at the surface). Typically, for transition metals the classical skin depth in rf range (10 GHz) is in the range of few microns. For thick magnetic films, Ament and Rado [17] showed that the eddy currents lead to a finite FMR linewidth even in the absence of damping. This effect is called as exchange conductivity mechanism, and in such case, the linewidth is proportional to $(A\sigma)^{0.5}$ where A is the exchange stiffness constant and σ is the conductivity. For samples that are thinner than the skin depth,

the contribution from eddy currents to magnetization dynamics equation is calculated by integrating the Maxwell's equation over the film thickness d. The calculation yields eddy current contribution to Gilbert damping as expressed below:

$$\alpha_{\text{eddy}} = \frac{1}{6} M_s \gamma \left(\frac{4\pi}{c}\right)^2 \sigma d^2 \tag{3.3}$$

where γ is the gyromagnetic ratio. As the eddy current damping is proportional to
d^2, hence, it is significantly reduced for thin-film samples. For the case of iron, the
eddy currents become important by a thickness of just about 25 nm, whereas for
nickel, thicknesses of up to 100 nm have quite small damping contributions from
eddy currents [18, 19].

3.2 Extrinsic Contribution

In this section, we discuss the extrinsic contribution to damping, which is observed
in almost all practical samples. Such contributions are of great interest because they
are subject to control through sample preparation. Careful sample preparation can
thus substantially reduce extrinsic sources of magnetic damping to produce extremely narrow FMR linewidth. Conversely, in selected instances, one may wish to
see magnetization precession to be more heavily damped. As an example of the
latter case, it is desirable to suppress 'ringing' of magnetization after reversal in
certain devices. Thus, identification of specific extrinsic mechanisms of magnetic
damping is of technological importance. In general, the extrinsic damping is known
to result from sample inhomogeneities. Broadly, various effects are characterized as
inhomogeneities. To name a few, local defects, and non-uniformity of sample
thickness, variations in surface anisotropies associated with step edges, differences
in the magnetocrystalline anisotropy in local regions, lattice strains due to the
substrate imperfections in thin films, etc., are considered as inhomogeneities [20–
22]. To treat the sample inhomogeneities accurately is an extremely challenging
problem. In order to understand the inhomogeneities, it has been characterized as
strong inhomogeneities and weak inhomogeneities. The strong inhomogeneities refer
to the case when the inhomogeneous fields are large compared to the intrinsic
exchange and dipolar fields, whereas the weak inhomogeneities implies that the
inhomogeneous field is weak compared to the exchange and dipolar fields. The
presence of strong inhomogeneities induces mixing of the uniform mode with the
non-uniform modes, causing decoherence of the uniform mode. In such a case,
different areas of the sample have negligibly small interaction with each other
which subsequently leads to large number of local resonance fields causing additional broadening of the FMR linewidth. Contrary to above, when inhomogeneities
are weak the magnetization may precesses nearly uniformly. Below, we discuss
further in detail the mechanisms related to extrinsic contribution to damping.

3.2.1 Two-Magnon Scattering

The concept of two-magnon scattering was proposed few decades ago in 1960 by Sparks et al. [23] to explain the FMR linewidth broadening in an important magnetic insulator yttrium iron garnet (YIG). Subsequently, this concept was invoked for understanding the extrinsic contribution to damping in ferrite materials. In 1967, Patton et al. [24] used this concept to explain the damping in metallic films. Arias and Mills, in 1999 [25], introduced two-magnon scattering theory that can be applied to ultrathin films. In two-magnon scattering model of extrinsic damping, a uniformly precessing mode ($q = 0$) scatters into non-uniform precessing mode ($q \neq 0$). Following the conservation of energy, the resonant mode can scatter into spin-waves oscillating at same frequency, whereas, due to loss of translational invariance, the momentum conservation does not take place. It is important to emphasize here that overall in this process the magnetic excitations do not disappear, rather it pumps the magnetic energy into other modes which leads to dephasing of the magnetic resonant mode. In order to reach equilibrium, the magnetic energy has to be dissipated to the lattice by damping which implies that the two-magnon scattering mechanism is similar to the mode–mode coupling in small lateral geometries [26, 27].

The two-magnon scattering matrix is proportional to the components of Fourier transform of the magnetic inhomogeneities as expressed in Eq. 3.4

$$A(\mathbf{q}) = \int d\mathbf{r} \Delta U(\mathbf{r}) e^{-i\mathbf{q}\cdot\mathbf{r}} \tag{3.4}$$

where $U(\mathbf{r})$ corresponds to local anisotropy energy. The magnon momentum is not conserved in two-magnon scattering process due to sample inhomogeneities (loss of translational invariance), however the energy is conserved. For the case of ultrathin films, the magnon $\mathbf{q}$ vectors are confined to the film plane. With increasing q, the magnon energy with $\mathbf{q}$ parallel to the saturation magnetization decreases its energy in case of in-plane orientation. Eventually at $q = q_0$, the magnon energy crosses the energy of the homogeneous mode due to the effective exchange field. From this, it implies that the magnon with the wave vector $\mathbf{q}_0$ is degenerate with the homogeneous mode and it can participate in two-magnon scattering. With increasing angle ψ between the $\mathbf{q}$ vector and the saturation magnetization, value of q decreases and no degenerate mode remains available for the angle ψ larger than

$$\psi_{max} = arc \sin\left(\frac{H}{H + 4\pi M_{eff}}\right)^{1/2} \tag{3.5}$$

where H is the field at FMR and $4\pi M_{eff}$ is the effective demagnetizing field perpendicular to the film surface.

In the study of Arias and Mills [25], the role of two-magnon scattering in ultrathin films was addressed in the parallel configuration. Specifically, it was shown that the lateral variation in the perpendicular uniaxial interface anisotropy field originating from the interface roughness is the predominant source of the two-magnon scattering. Furthermore, Arias and Mills [25] derived the in-plane rf-susceptibility equation for magnetic thin films and showed through their calculation that it can be expressed as

$$\chi_{//} = \frac{M_S B}{B_{\text{eff}} H - \left(\frac{\omega}{\gamma}\right)^2 + i(H+B)\frac{\alpha\omega}{\gamma} + [\text{Re}(R) + i\text{Im}(R)]} \quad (3.6)$$

where $B_{\text{eff}} = H + 4\pi M_{\text{eff}}$ is the effective induction, R is the mass operator corresponding to two-magnon scattering and in-plane anisotropies are neglected for simplicity. It is important to point out here that the real part $\text{Re}(R)$ gives rise to a shift in the FMR field and the imaginary part $\text{Im}(R)$ results in additional damping. Another important condition is that the symmetry of magnetic inhomogeneities needs to be satisfied by $\text{Re}(R)$ and $\text{Im}(R)$. In case of a rectangular network of misfit dislocations, additional two- and four-fold anisotropies affecting both the FMR field and magnetic damping are expected to be present. Furthermore, note that both Re (R) and $\text{Im}(R)$ are dynamic effects and in general dependent on the microwave frequency. Following the concept of Fourier components of magnetic inhomogeneities, the strength of two-magnon scattering in damping as a function of the φ_M (angle of the magnetization with respect to the in-plane crystallographic axis) is expressed as

$$\text{Im}[R(\varphi_M)] \sim \int I(\mathbf{q})\delta(\omega - \omega_q)d\mathbf{q}^3 = 2 \int_{-\psi_{\max}}^{\psi_{\max}} I(q_0, \varphi, \varphi_M)\frac{q_0 d\psi}{\frac{\partial\omega}{\partial q}(q_0, \psi)} \quad (3.7)$$

where $I(\mathbf{q})$ denotes the effective intensity of two-magnon scattering, and $\varphi = \varphi_M + \psi$ is defined as the angle of the $\mathbf{q}$ vector with respect to the defect axis. The magnon group velocity given by $\partial\omega/\partial q$ (q_0, ψ) is proportional to the strength of the dipolar and exchange fields. Importantly, this represents the dipole exchange narrowing of local inhomogeneities. One can interpret from Eq. (3.7) that the number of degenerate magnons is proportional to ω^2 as $\omega \to 0$, and consequently the two-magnon scattering eventually decreases linearly to zero with decreasing microwave frequency. Overall, in this mechanism magnons are coupled together by the scattering matrix which is given by components of the Fourier transform of magnetic defects. This results in coupled linear equations which can be solved exactly by evaluating eigenmodes of the system. The resonant mode is then given by superposition of the eigenmodes, and the FMR linewidth is due to the spread of the eigen frequencies (dispersion relationship). It is not necessary that the extrinsic FMR linewidth broadening is always due to two-magnon scattering. In order to confirm the presence of two-magnon scattering contribution, it is also required that by using FMR measurements in perpendicular magnetization configuration, ΔH as a function of microwave frequency should be described by Gilbert damping.

In the interpretation of extrinsic contribution to damping, the finite value of $H(0)$ in the perpendicular magnetization configuration may exist even if the two-magnon scattering is present in the system. However, the long-wavelength variations are not a part of two-magnon scattering and can be present in the perpendicular configuration. We thus emphasize here that the interpretation of two-magnon scattering is not simple and requires a careful evaluation.

We now discuss some of the experimental observations of two-magnon scattering with a particular focus on the study of Woltersdorf et al. [26, 28]. In their work, the signature of two-magnon scattering originating from structural defects in ultrathin film multilayers of epitaxial Au/Pd/Fe and Au/Fe/Pd/Fe grown on GaAs (001) are investigated in detail. The structural defects are mainly present due to the lattice misfit between individual layers (Pd has a lattice mismatch of 4.4% with respect to Fe and 4.9% with respect to Au). Thus, the relaxation of lattice strain affects these samples by creating a self-assembled network of misfit dislocations when the Pd layer is sufficiently thick. Specifically, using structural characterization, the network of misfit dislocations was found to be oriented along the (100) crystallographic axes of Fe. The presence of the network of misfit dislocations is expected to result in strong extrinsic damping. In this study, they found that for n Pd/Fe/GaAs(001) structures, with the number of Pd atomic layers $n \geq 130$, the FMR linewidth depends strongly on the angle (ϕ_M) between the magnetization and the crystallographic <100> Fe axes. By investigating the FMR linewidth as a function of ϕ_M, the in-plane easy axis (along <100> Fe) and in-plane hard axis (along <110> Fe) were found. In this study, subsequently they measured the FMR linewidth as a function of frequency along these directions. In Fig. 3.1 we reproduce the FMR linewidth ΔH versus frequency plot as observed in the experiment of Woltersdorf et al. [26] along <100> Fe and <110> Fe. Note that along <110> Fe (symbols represented using filled star), the FMR linewidth between 10 and 73 GHz is almost linearly dependent on the microwave frequency along with an offset ΔH (0) = 50 Oe at zero-frequency value. In this figure, the plot with open star represents the intrinsic Gilbert damping obtained in samples and the slope for this linear plot is similar to the one with <110> Fe. The presence of 100 Pd(001) overlayer allows the contribution from spin pumping to the Gilbert damping [9] (for details of spin pumping see Chap. 7). Also in this study it was found that for 30 ML thick Fe film, the <100> Fe direction is neither easy axis nor hard axis. Due to this, dragging of magnetization behind the external field occurs which leads to additional linewidth broadening in FMR. In particular, at low frequencies the dragging contribution is present as the FMR fields are comparable to the in-plane uniaxial anisotropy field and in the above case the dragging is absent above 24 GHz [26]. Interestingly, for the <100> Fe orientations (filled circles), ΔH has clearly nonlinear dependence on the microwave frequency contrary to expectations for Gilbert damping. Also, between 36 GHz and 73 GHz the slope of the FMR linewidth is similar to that due to intrinsic damping. It is important to notice here that for the <100> Fe orientations, $\Delta H(0)$ = 160 Oe, which is significantly larger than the value for <110> Fe orientation. Furthermore, a noticeable feature in Fig. 3.1 is observed below 36 GHz where the frequency dependence of ΔH nonlinearly drops

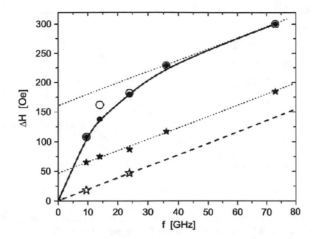

Fig. 3.1 The frequency dependence of the FMR linewidth for the 200 Pd/**30 Fe**/GaAs(001) structure along the $<110>_{Fe}$ (★) and $<100>_{Fe}$ (●) directions, respectively. (○) show ΔH along $<100>_{Fe}$ before the dragging contribution to the linewidth was removed (see further details in the text). The purpose of the *solid line* spline fit is to guide the reader's eye. The *dashed line* shows the frequency dependence of the intrinsic FMR linewidth (Gilbert damping) obtained by using the 100 Pd/**30 Fe**/GaAs(001) sample with no magnetic defects in Fe. The spin pumping contribution in 100 Pd/**30 Fe**/GaAs(001) to ΔH was already saturated Ref. [9]. ☆ symbols on the *dashed line* show the FMR linewidth in the perpendicular configuration at 10 and 24 GHz for the 200 Pd/**30 Fe**/ GaAs(001) sample. Note that ☆ are right on the *dashed line* indicating that in the perpendicular configuration the FMR linewidth ΔH is only given by the Gilbert damping ($\alpha = 0.006$) with no zero-frequency offset [$\Delta H(0)_\perp = 0$]. The *dotted lines* indicate the range of microwave frequencies where the slope of $\Delta H(f)$ is close to that expected from the Gilbert damping. Note that the *dotted lines* have zero-frequency offsets. *Reprinted with permission from Ref.* [26]. *Copyright 2004 by the American Physical Society*

to zero at small value of microwave frequencies. Most importantly, the frequency dependence for the <100> Fe orientation is consistent with the calculations of two-magnon scattering in ultrathin films as proposed by Arias and Mills [25].

To confirm the presence of two-magnon scattering convincingly it is required to investigate the FMR linewidth as a function of the angle θ_M between the magnetization M and the sample surface by applying the bias field away from the sample surface. In Fig. 3.2, we reproduce the dependence of the FMR linewidth on the angle θ_H between the bias magnetic field and the sample plane from Woltersdrof et al. [26]. These results have shown that the damping decreases significantly when the magnetization is almost perpendicular to the film surface. Interestingly, in this study, they found that the measured ΔH in the perpendicular configuration at 10 and 24 GHz correspond to the same value as intrinsic damping of the film. From the FMR peak narrowing shown in Fig. 3.2, and the absence of $\Delta H(0)$ in the perpendicular configuration, the presence of two-magnon scattering in the Pd/Fe/GaAs (001) structures was inferred in this study.

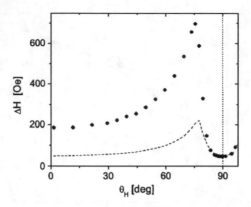

Fig. 3.2 Measured ferromagnetic resonance linewidth ΔH_m as a function of θ_H at 24 GHz. The *dots* represent the measured data and the *dashed line* represents the FMR linewidth $\Delta H_G(\theta)$ that was calculated using the Gilbert damping from the perpendicular configuration. The peak in the FMR linewidth for $\theta = 78°$ is caused by dragging the magnetization behind the applied field. *Reprinted with permission from Ref.* [26]. *Copyright 2004 by the American Physical Society*

Furthermore, from out-of-plane FMR measurements in this study, additional features confirming the presence of two-magnon scattering were noticed. In Fig. 3.2, by taking intrinsic value of the Gilbert damping $G_{int} = 1.4 \times 10^8$ s^{-1}, the calculated dependence of the FMR linewidth as a function of θ_H is shown by the dashed line. Note that the calculated increase for the intermediate angles is due to the dragging of the magnetic moment behind the bias field H. Confirmation of the presence of two-magnon scattering is the noticeable difference between the measured FMR linewidth and that expected for the intrinsic damping ΔH_{ext}. The strength of two-magnon scattering as a function of the angle θ_H is given by the adjusted frequency linewidth, $\Delta \omega$,

$$\frac{\Delta \omega}{\gamma} = \left(\frac{d\omega}{dH}\right) \Delta H_{ext}(\theta_H) \tag{3.8}$$

where $d\omega/dH = [\omega(H + \Delta H, \theta + \Delta\theta) - \omega(H, \theta)]/\Delta H$ using the FMR condition for the resonance frequency which includes the in-plane and out-of-plane magnetic anisotropies [29]. In order to calculate $d\omega/dH$, it is required to choose $\Delta\omega$ and evaluate the appropriate change in ΔH and $\Delta\theta$ which satisfy the resonance condition for the out-of-plane configuration.

In Fig. 3.3, the variation of $\Delta\omega/\gamma$ calculated from the measured FMR linewidth data is reproduced from Woltersdrof et al. Interestingly from Fig. 3.3 one can note that the extrinsic contribution to the frequency linewidth $\Delta\omega/\gamma$ is nearly independent of θ_H, however, it decreases sharply when the magnetization angle $\theta_M \sim \pi/4$. Also from this figure it may be noted that $\Delta\omega/\gamma$ and ψ_{max} follow well each other as a function of θ_H (cf. dashed line in Fig. 3.3). The observed proportionality between $\Delta\omega/\gamma$ and $\psi_{max}(\theta_H)$ provides most convincing evidence that this extrinsic damping

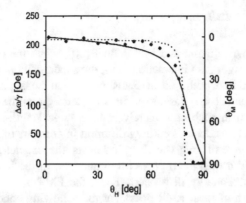

Fig. 3.3 (●) represent the adjusted frequency FMR linewidth $\Delta\omega/\gamma$ from the extrinsic contribution as a function of θ_H at 24 GHz. The *solid line* shows the angle of magnetization θ_M as a function of θ_H. The *dashed line* shows the critical angle ψ_{max} as a function of θ_H. Notice that ψ_{max} describes the angular dependence of $\Delta\omega/\gamma$ quite well. ψ_{max} was scaled in order to compare it with $\Delta\omega/\gamma$. *Reprinted with permission from Ref.* [26]. *Copyright 2004 by the American Physical Society*

is due to two-magnon scattering. It may be also inferred from this plot that the effective scattering matrix I $(\mathbf{q}, \theta_M)$ is weakly dependent on θ_M. It is important to mention here that the intensity of two-magnon scattering can have an explicit dependence on the direction of the magnetization with respect to the axes of magnetic defects, and it may be at the origin of a strong in-plane angular dependence of the two-magnon scattering in self-assembled network of misfit dislocations.

In various systems having contribution from extrinsic damping, the zero frequency offset can be observed. For the cases of metallic thin films and multilayers the FMR linewidth, ΔH in low microwave frequency range (cf. between 10 and 36 GHz in Fig. 3.1) is in general described by,

$$\Delta H(\omega) = \Delta H(0) + 1.16 \frac{\omega}{\gamma} \frac{G_{eff}}{\gamma M_S} \tag{3.9}$$

where $\Delta H(0)$ is the zero-frequency offset originating from the extrinsic contribution to the FMR linewidth. The effective Gilbert damping (G_{eff}) contains both the intrinsic (G_{int}) and extrinsic (G_{ext}) contributions. As the microwave frequency approaches zero the contribution of two-magnon scattering becomes zero however other extrinsic contribution may still be present in $\Delta H(0)$.

3.2.2 Magnetic Inhomogeneity

As proposed by Twisselmann and McMichael [30], when the characteristic inhomogeneity field is larger than interaction field, the superposition of local resonances obscures the FMR linewidth. Local FMR peaks can be superposed in case of long-wavelength (small q) variations in magnetic properties of the sample. The FMR linewidth in such case merely reflects large length-scale sample inhomogeneities, and this scenario is entirely different in comparison with two-magnon scattering. In the range of long-wavelength defects, the magnetostatic contribution $2\pi M_s q d$ (d is the film thickness) plays important role in the interaction field thus affecting the magnon-energy dispersion. Thus, the FMR spectrum is given by a simple superposition of local FMR peaks for the following condition:

$$H_P D \geq 3\pi M_s d \qquad\qquad (3.10)$$

where H_P is the root-mean-square value of random variations of a local anisotropy field satisfying a Gaussian distribution and D is the corresponding average grain size. Presence of these magnetic inhomogeneity may cause a finite $\Delta H(0)$ at zero frequency.

3.3 Theoretical Model for Damping

After providing a glimpse for various contributions to the Gilbert damping, we now discuss few important theoretical models for it. These models were developed during 1970–1980, and these are extremely important to obtain a general picture of the damping mechanism. It is mostly argued that the leading mechanism of damping in metals is the spin–orbit interaction with the itinerant nature of electrons. In the description provided in subsections below, the mathematical expressions are used to discuss the parameters needed for understanding intrinsic damping without getting into complicated real calculations [4, 10].

3.3.1 s-d-Exchange Relaxation

In 1976, Kambersky [3] showed that the intrinsic damping in ferromagnetic metals can be treated by considering spin–orbit coupling Hamiltonian. In this study, the main idea was related to the incoherent scattering of itinerant conduction electrons by phonons and magnons which contributes to Gilbert damping in metals. It was proposed that the loss of energy and angular momentum from localized d-electrons is mediated by the exchange interaction to the itinerant s-electrons. The spin transfer from d to s states is accompanied by the relaxation of the s-electrons' spins to the lattice through an incoherent spin-flip process due to the spin–orbit coupling. The particle representation of s–d exchange interaction Hamiltonian for the rf-components of the magnetization

$$H_{sd} = \sqrt{\frac{2S}{N}} \sum_{k} J(q) a_{k,\uparrow} a_{k+q,\downarrow}^{\dagger} b_q + h.c. \qquad (3.11)$$

where S and N represent spins and number of atomic sites. Here, creation and annihilation of electron are represented by $a^{\dagger}$ and a, whereas for magnon, the same is denoted by $b^{\dagger}$ and b. $\uparrow$ and $\downarrow$ denote majority and minority electrons, respectively. The magnons and itinerant electrons are coherently scattered by the s–d exchange interaction which results in creation (first term in the right hand side of the equation) and annihilation (Hermitian conjugate, h.c.) of electron–hole pairs as indicated by Eq. 3.11. During the scattering with magnons, the itinerant electron spin flips and the total angular momentum in the s–d exchange interaction remains conserved as shown in Fig. 3.4. The s–d exchange interaction on its own leads to renormalized spectroscopic splitting factor. Thus, it is important to disrupt the coherent scattering of magnons with itinerant electrons by incoherent scattering with other excitations, for example, incoherent scattering by thermally excited phonons and magnons or by strong spin–orbit scattering center. In order to account for incoherent scattering, a finite lifetime τ_{eff} (effective lifetime) is defined for electron hole pair excitation in Eq. 3.12

$$\delta\varepsilon = \varepsilon_{k+q}^{\downarrow} - \varepsilon_k^{\uparrow} + i \frac{\hbar}{\tau_{eff}} \qquad (3.12)$$

In general, τ_{eff} is defined by spin-flip time τ_{flip} of electron–hole pair. Due to the spin–orbit interaction, which creates a nonzero scattering probability between the two-spin states, the rate of spin flip gets modified, and it is given by $\tau_{flip} = \tau_{orb}/(g-2)^2$, where τ_{orb} is the orbital relaxation time and g is the factor which depends on the ratio of the spin and the orbital momentum. A better estimate of τ_{flip} may be

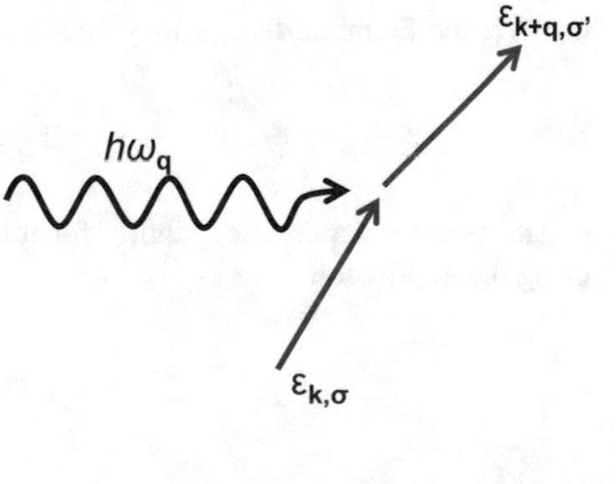

Fig. 3.4 Schematic depicting collision of spin wave (energy $h\omega_q$) with itinerant electron with energy $\varepsilon_{k\sigma}$ and creating itinerant electron of energy $\varepsilon_{k+q,\sigma'}$. σ and σ' indicate spin states

obtained by evaluating the spin-diffusion length l_{sd}, and these are related via Eq. 3.13.

$$\tau_{flip} = \frac{6l_{sd}^2}{v_F} \frac{\lambda_{\uparrow} + \lambda_{\downarrow}}{\lambda_{\uparrow}\lambda_{\downarrow}} \qquad (3.13)$$

where v_F is the Fermi velocity, $\lambda_{\uparrow}$ and $\lambda_{\downarrow}$ are the electron momentum free paths for spin-up and spin-down electrons.

The rf-susceptibility as discussed in expression 3.6 may be calculated using the Kubo Green's function formalism under random phase approximation (RPA) [4]. In this equation, the imaginary part in the denominator of circularly polarized rf-susceptibility can be given by effective damping field (Eq. 3.14):

$$\alpha_{s-d} \frac{\omega}{\gamma} = \frac{2\langle S \rangle}{Ng\mu_B} \sum_k |J(q)|^2 \left(n_{k+q}^{\downarrow} - n_k^{\uparrow} \right) \delta\left(\hbar\omega_q + \varepsilon_k^{\uparrow} - \varepsilon_{k+q}^{\downarrow} \right) \qquad (3.14)$$

where $<S>$ is the reduced spin given by $<S> = S\,(M_S(T)/M_S(0))$, n the occupation number, δ represents the Dirac delta function, and the summation is performed over all available states on the Fermi surface. The incoherent scattering of electron–hole pair excitation causes broadening of delta function in Eq. 3.14 into Lorentzian function as expressed below in Eq. 3.15,

$$\delta\left(\hbar\omega_q + \varepsilon_k^{\uparrow} - \varepsilon_{k+q}^{\downarrow} \right) \sim \frac{\hbar/\tau_{flip}}{\left(\hbar\omega_q + \varepsilon_k^{\uparrow} - \varepsilon_{k+q}^{\downarrow} \right)^2 + \left(\hbar/\tau_{flip} \right)^2} \qquad (3.15)$$

The difference in the occupation numbers $\Delta n = \left(n_{k+q}^{\downarrow} - n_k^{\uparrow} \right) = \delta(\varepsilon_k - \varepsilon_F)\hbar\omega_q$ where ε_F is the Fermi energy. The presence of delta function takes into account only the relaxation process of electrons present at Fermi level, and $\hbar\omega_q$ refers to resonant magnon energy involved in the scattering process. Considering the case of ferromagnetic resonance experiments which are mainly sensitive to the uniform precession mode ($q \sim 0$) and taking into account the fact that the change in the energy for spin-flip electron–hole pair excitation is dominated by exchange energy, one can write $\left(\varepsilon_{k+q}^{\downarrow} - \varepsilon_k^{\uparrow} \right) = -2\langle S \rangle J(0)$. By using $N<S>g\mu_B = M_S(T)$ and after integrating over the Fermi surface, it may be shown that the Gilbert damping parameter is

$$\alpha_{s-d} = \frac{\chi_P}{M_S\gamma} \frac{1}{\tau_{flip}} \qquad (3.16)$$

where χ_P is the Pauli susceptibility for itinerant electrons which can be estimated using the expression

$$\chi_P = \left(\frac{\hbar\gamma}{2\pi}\right)^2 \int k^2 dk \delta(\varepsilon_k - \varepsilon_F) = \mu_B^2 N(E_F) \tag{3.17}$$

where $N(E_F)$ is the density of states at Fermi level. The value of χ_P for 3d transition metals is in the range of 3×10^{-6}– 9×10^{-6} [31]. Overall as a consequence of Eq. 3.16, in the absence of thermally excited magnons, it has been interpreted that the s–d relaxation is inversely proportional to spin-flip scattering time and it scales with the resistivity. It is worth to mention here that Ingvarsson et al. [32] had shown recently that the s–d scattering successfully accounts for the damping in $Ni_{81}Fe_{19}$ films, and it scales with sample resistivity. Interestingly, in case of pure 3d ferromagnets such as Fe, Co, and Ni, spin-flip relaxation time is significantly large, and in turn, s–d scattering fails to account for Gilbert damping parameter.

By taking into account of scattering without spin-flip, Kambersky [3] showed that the intrinsic damping in metallic ferromagnets can be treated by spin–orbit interaction Hamiltonian as expressed in Eq. 3.18.

$$H^{SO} = \frac{1}{2}\frac{\sqrt{2S}}{N}\xi\sum_k\sum_{\mu,\upsilon,\sigma}\langle\mu|L^+|\upsilon\rangle c^\dagger_{\upsilon,k+q,\sigma}c_{\mu,k,\sigma}b_q + h.c. \tag{3.18}$$

where the indices μ and υ represent the projected local orbital of Bloch states, $L^\pm = L_x \pm iL_y$ are right-handed and left-handed components of the atomic site transverse angular momentum L, $c_{\mu,k,\sigma}$, and $c^\dagger_{\upsilon,k+q,\sigma}$ annihilation, and creation of electrons in the appropriate Bloch states with spin σ, b_q represents the annihilation of the spin wave with wave vector q. In order to simplify the discussion, the dependence of the matrix elements $\langle\mu|L^+|\upsilon\rangle$ on the wave vectors is neglected. Following Kubo Green's function formalism under RPA the rf-susceptibility can be calculated and subsequently damping may be denoted as Eq. 3.19 [4]

$$\alpha_{so}\frac{\omega}{\gamma} = \frac{\langle S\rangle^2}{2M_s}\xi^2\left(\frac{1}{2\pi}\right)^3\int d^3k \sum_{\mu,\upsilon,\sigma}\langle\upsilon|L^+|\mu\rangle\langle\mu|L^-|\upsilon\rangle$$

$$\times \delta(\varepsilon_{\mu,k,\sigma} - \varepsilon_F)\hbar\omega\frac{\hbar/\tau_m}{(\hbar\omega + \varepsilon_{\mu,k,\sigma} - \varepsilon_{\upsilon,k+q,\sigma})^2 + (\hbar/\tau_m)^2} \tag{3.19}$$

where τ_m is the momentum relaxation time. In this case, τ_m is used due to absence of spin-flip scattering. Based on the above discussion, we discuss two specific cases, namely (a) intraband transition and (b) interband transition.

(a) **Intraband Transition** ($\mu = \upsilon$): In case of small-wave-vector spin waves ($q \ll k_F$), the electron energy balance $\hbar\omega + \varepsilon_{\mu,k,\sigma} - \varepsilon_{\mu,k+q,\sigma} = \hbar\omega - (\hbar^2/2m)$ ($2kq + q^2$) in the denominator of Eq. 3.19 is significantly smaller than $\hbar/\tau_m$. In a well-crystalline system, this limit is satisfied at higher temperature (above cryogenic temperature). The Gilbert damping for such a case may be estimated by integrating over the Fermi surface, and it can be expressed as Eq. 3.20

$$\alpha_{so}^{intra} \simeq \frac{\langle S \rangle^2}{M_s \gamma} \left(\frac{\xi}{\hbar} \right)^2 \left(\sum_{\mu} \chi_P^{\mu} \langle \mu | L^+ | \mu \rangle \langle \mu | L^- | \mu \rangle \right) \tau_m \tag{3.20}$$

Here, χ_P^{μ} denotes the Pauli susceptibility of those states which participate in the intraband transition and satisfy $|\hbar\omega - (\hbar^2/2m)(2kq + q^2)| < < \hbar/\tau_m$. It is interesting to note that in this limit, the Gilbert damping is proportional to τ_m and it scales with conductivity.

(b) Interband transition ($\mu \neq v$): The energy gap $\Delta\varepsilon_{v,\mu}$ accompanying the interband transition and the electron–hole pair energy can be dominated by these gaps. For the gaps which are larger than the momentum relaxation frequency $\hbar/\tau_m$, the Gilbert damping may be expressed by

$$\alpha_{so}^{inter} \simeq \frac{\langle S \rangle^2}{M_s \gamma} \sum_{\mu} \chi_P^{\mu} (\Delta g_{\mu})^2 \frac{1}{\tau_m} \tag{3.21}$$

where

$$\Delta g_{\mu} = 4\xi \sum_{v} \langle \mu | L_x | v \rangle \langle v | L_x | \mu \rangle / \Delta\varepsilon_{v,\mu} \tag{3.22}$$

which determines the contribution of the spin–orbit interaction to the g-factor. Under this assumption, the Gilbert damping parameter is proportional to $1/\tau_m$, and it scales with resistivity. In general, a large distribution of energy gap describes the temperature dependence of the Gilbert damping. Thus, only at low temperatures, the interband damping scales with the resistivity, whereas at high temperatures, the relaxation rate $\hbar/\tau_m$ becomes comparable to the energy gaps, $\Delta\varepsilon_{\mu,v}$ resulting in gradual saturation.

In classical picture, the s–d exchange interaction can be described by considering two precessing magnetic moments corresponding to the d-localized and itinerant s-electrons which are mutually coupled by the s–d exchange field. In case the damping is absent, the low-energy excitation corresponds to a parallel alignment of the magnetic moments coherently precessing in phase. As the spin mean free path of the itinerant electrons has finite value, hence its equation of motion includes the spin relaxation toward the instantaneous effective field given by $\frac{1}{\tau_{flip}\gamma}(m - \chi_P h_{eff})$ where the exchange field between itinerant and localized electrons is incorporated in h_{eff}. Overall, this results in a phase lag between d and s precessing moments and consequently to damping. The contribution to the Gilbert damping is proportional to the rate of spin flip, i.e. $1/\tau_{flip}$. Thus, the spin–orbit coupling plays crucial role in defining the damping in metals.

3.3.2   Fermi Surface Breathing

It is well known that in the ferromagnetic metals, the shape of the Fermi surface changes when the magnetization direction is changed. As the precession of magnetization evolves in time and space, the Fermi surface also gets periodically distorted. Hence, the uniform precession results in a periodical variation of the Fermi surface due to spin–orbit coupling (often called breathing Fermi surface) [33]. As in this process, itinerant electrons are involved in changing the Fermi surface, thus the overall process is dissipative and it introduces a dephasing between the magnetization precession and the periodic variation. Thus, for the scattered itinerant electrons, relaxation occurs due to the repopulation of the Fermi surface, and overall in this process, a phase lag develops between the Fermi surface distortion and the precessing magnetization. This relaxation corresponds to Gilbert-like contribution to the damping and primarily depends on the strength of the spin–orbit coupling. In this model, the damping matrix is determined by the dependence of the effective single-electron energies on the modification of the magnetic configuration, and by the electronic scattering processes which transfer angular momentum from the electronic spin system to the lattice. The latter process is mediated by the spin–orbit coupling. Detailed mathematical calculations for breathing Fermi surface using a sophisticated treatment of spin–orbit interaction can be found in ref. [3, 8]. In 2002, Kunes and Kambersky [34, 35] carried out first principles electronic band calculations of the breathing Fermi surface Gilbert damping in Ni, Co, and Fe. In this study, reasonably good quantitative agreement between the experimental data and calculations was found.

In 2005, Steiauf and Fahnle discussed further modification in the original breathing Fermi surface by using ab initio density-functional theory, where the single-electron functions are used to describe electron scattering [36]. In this theory, the magnetization is described by the effective single particle theory which uses a wave function $\psi_{j,k}$ for each electron in a band with band index j and wave vector k. Subsequently, following density functional theory, the band structure energy is calculated by summation of single-electron energy ϵ_{jk}. The detailed calculation showed that the single-electron energy depends on the orientation of the atomic magnetic moment $e_i(t)$. Overall, a small change in the $e_i(t)$ causes modification in the Fermi surface [37]. Based on the breathing, Fermi surface model of Gilbert damping and on the Elliott-Yafet relation for the spin-relaxation time, in 2010, Fahnle et al. [38] postulated the relation between Gilbert damping and ultrafast laser-induced magnetization dynamics. In their calculation, they derived a relation between the conductivity like contribution to the Gilbert damping at low temperatures and the demagnetization time τ_{demag} for ultrafast laser-induced demagnetization at low laser fluences. It was further concluded that the same type of spin-dependent electron scattering mechanisms are relevant for Gilbert damping and demagnetization time. Though earlier attempts of providing a unified theory for damping and ultrafast demagnetization was made by Koopmans et al. using microscopic model [39], the calculations of Fahnle et al. by taking into account breathing Fermi surface model successfully explained various experimental results [38, 40]. We wish to emphasize here that we have intentionally limited our discussion in this chapter for ferromagnetic thin films and mostly with uniform magnetization precession where local damping is relevant to consider. The Gilbert damping model in the local and time-independent approximation shows a linear dependence of the FMR linewidth on resonance frequency. It is worth to mention here that in spatially dependent magnetization textures, the non-local character of the damping may be significant. The non-uniform magnetization precession can cause space-dependent Gilbert damping referred as non-local-damping, which is driven by magnetization dynamics at other sites [41–43]. Importantly, the anisotropic and non-local character of the magnetic damping is defined by damping tensor α_{ij}, and some of the recent theoretical studies have derived it within the Kamberský model. Later in Chap. 7, we will discuss other mechanisms, e.g., spin pumping and interfacial d–d hybridization, which actively govern the damping in ferromagnet/non-magnet bilayer systems [44–48].

References

1. Gilbert TL (1955) A lagrangian formulation of the gyromagnetic equation of the magnetic field. Phys Rev 100:1243
2. Gilbert TL (2004) A phenomenological theory of damping in ferromagnetic materials. IEEE Trans Magn 40(6):3443–3449. doi:10.1109/tmag.2004.836740
3. Kamberský V (1976) On ferromagnetic resonance damping in metals. Czech J Phys B 26 (12):1366–1383. doi:10.1007/bf01587621
4. Heinrich B, Bland JAC (2005) Spin relaxation in magnetic metallic layers and multilayers. In: Bland JAC (ed) Ultrathin magnetic structures: fundamentals of nanomagnetism, vol 3. Springer, New York
5. Landau L, Lifshitz E (1935) On the theory of the dispersion of magnetic permeability in ferromagnetic bodies. Phys Z Sowjetunion 8(153):101–114
6. Kittel C (1948) On the theory of ferromagnetic resonance absorption. Phys Rev 73(2):155–161. doi:10.1103/PhysRev.73.155
7. Heinrich B, Cochran JF (1993) Ultrathin metallic magnetic films: magnetic anisotropies and exchange interactions. Adv Phys 42(5):523–639. doi:10.1080/00018739300101524
8. Kamberský V (1984) FMR linewidth and disorder in metals. Czech J Phys B 34(10):1111–1124. doi:10.1007/bf01590106
9. Foros J, Woltersdorf G, Heinrich B, Brataas A (2005) Scattering of spin current injected in Pd (001). J Appl Phys 97(10):10A714. doi:10.1063/1.1853131
10. Heinrich B, Urban R, Woltersdorf G (2002) Magnetic relaxations in metallic multilayers. IEEE Trans Magn 38(5):2496–2501. doi:10.1109/tmag.2002.801906
11. van Kampen M, Jozsa C, Kohlhepp JT, LeClair P, Lagae L, de Jonge WJM, Koopmans B (2002) All-optical probe of coherent spin waves. Phys Rev Lett 88(22):227201. doi:10.1103/PhysRevLett.88.227201
12. Barman A, Wang S, Maas JD, Hawkins AR, Kwon S, Liddle A, Bokor J, Schmidt H (2006) Magneto-optical observation of picosecond dynamics of single nanomagnets. Nano Lett 6 (12):2939–2944. doi:10.1021/nl0623457
13. Gilmore K (2007) Precession damping in itinerant ferromagnets, Ph.D. Thesis, Chapter 3. Montana State University

14. He P, Ma X, Zhang JW, Zhao HB, Lüpke G, Shi Z, Zhou SM (2013) Quadratic scaling of intrinsic Gilbert damping with spin-orbital coupling in L1$_0$ FePdPt films: experiments and ab initio calculations. Phys Rev Lett 110(7):077203. doi:10.1103/PhysRevLett.110.077203

15. Suhl H (1998) Theory of the magnetic damping constant. IEEE Trans Magn 34(4):1834–1838. doi:10.1109/20.706720

16. Heinrich B, Cochran JF, Myrtle K (1982) The exchange splitting of phonon assisted microwave transmission at 9.5 GHz. J Appl Phys 53(3):2092–2094. doi:10.1063/1.330708

17. Ament WS, Rado GT (1955) Electromagnetic effects of spin wave resonance in ferromagnetic metals. Phys Rev 97(6):1558–1566. doi:10.1103/PhysRev.97.1558

18. Cochran JF, Heinrich B, Arrott AS (1986) Ferromagnetic resonance in a system composed of a ferromagnetic substrate and an exchange-coupled thin ferromagnetic overlayer. Phys Rev B 34(11):7788–7801. doi:10.1103/PhysRevB.34.7788

19. Heinrich B, Fraitová D, Kamberský V (1967) The influence of s-d exchange on relaxation of magnons in metals. Phys Status Solidi (b) 23(2):501–507. doi:10.1002/pssb.19670230209

20. McMichael RD, Twisselmann DJ, Kunz A (2003) Localized ferromagnetic resonance in inhomogeneous thin films. Phys Rev Lett 90(22):227601. doi:10.1103/PhysRevLett.90.227601

21. McMichael RD, Twisselmann DJ, Bonevich JE, Chen AP, Egelhoff WFE Jr, Russek SE (2002) Ferromagnetic resonance mode interactions in periodically perturbed films. J Appl Phys 91(10):8647–8649. doi:10.1063/1.1456382

22. McMichael RD, Krivosik P (2004) Classical model of extrinsic ferromagnetic resonance linewidth in ultrathin films. IEEE Trans Magn 40(1):2–11. doi:10.1109/tmag.2003.821564

23. Sparks M, Loudon R, Kittel C (1961) Ferromagnetic relaxation. I. theory of the relaxation of the uniform precession and the degenerate spectrum in insulators at low temperatures. Phys Rev 122(3):791–803. doi:10.1103/PhysRev.122.791

24. Patton CE, Wilts CH, Humphrey FB (1967) Relaxation processes for ferromagnetic resonance in thin films. J Appl Phys 38(3):1358–1359. doi:10.1063/1.1709621

25. Arias R, Mills DL (1999) Extrinsic contributions to the ferromagnetic resonance response of ultrathin films. Phys Rev B 60(10):7395–7409. doi:10.1103/PhysRevB.60.7395

26. Woltersdorf G, Heinrich B (2004) Two-magnon scattering in a self-assembled nanoscale network of misfit dislocations. Phys Rev B 69(18):184417. doi:10.1103/PhysRevB.69.184417

27. Mills DL, Rezende S (eds) (2003) Spin damping in ultrathin magnetic films. Springer, Berlin

28. Woltersdorf G (2004) Spin-pumping and two magnon scattering in magnetic multilayers, Ph. D. Thesis, Chapter 2. Simon Fraser University

29. Hurben MJ, Patton CE (1998) Theory of two magnon scattering microwave relaxation and ferromagnetic resonance linewidth in magnetic thin films. J Appl Phys 83(8):4344–4365. doi:10.1063/1.367194

30. Twisselmann DJ, McMichael RD (2003) Intrinsic damping and intentional ferromagnetic resonance broadening in thin permalloy films. J Appl Phys 93(10):6903–6905. doi:10.1063/1.1543884

31. Kriessman CJ, Callen HB (1954) The magnetic susceptibility of the transition elements. Phys Rev 94(4):837–844. doi:10.1103/PhysRev.94.837

32. Ingvarsson S, Ritchie L, Liu XY, Xiao G, Slonczewski JC, Trouilloud PL, Koch RH (2002) Role of electron scattering in the magnetization relaxation of thin Ni$_{81}$Fe$_{19}$ films. Phys Rev B 66(21):214416. doi:10.1103/PhysRevB.66.214416

33. Kamberský V (1970) On the landau-lifshitz relaxation in ferromagnetic metals. Can J Phys 48(24):2906–2911. doi:10.1139/p70-361

34. Kuneš J, Kamberský V (2002) First-principles investigation of the damping of fast magnetization precession in ferromagnetic 3d metals. Phys Rev B 65(21):212411. doi:10.1103/PhysRevB.65.212411

35. Kuneš J, Kamberský V (2003) Erratum: first-principles investigation of the damping of fast magnetization precession in ferromagnetic 3d metals [Phys Rev B 65:212411 (2002)]. Phys Rev B 68(1):019901. doi:10.1103/PhysRevB.68.019901

36. Steiauf D, Fähnle M (2005) Damping of spin dynamics in nanostructures: An ab initio study. Phys Rev B 72(6):064450. doi:10.1103/PhysRevB.72.064450
37. Fahnle M, Steiauf D (2006) Breathing fermi surface model for noncollinear magnetization: a generalization of the Gilbert equation. Phys Rev B 73(18):184427. doi:10.1103/PhysRevB. 73.184427
38. Fahnle M, Seib J, Illg C (2010) Relating Gilbert damping and ultrafast laser-induced demagnetization. Phys Rev B 82(14):144405. doi:10.1103/PhysRevB.82.144405
39. Koopmans B, Ruigrok JJM, Longa FD, de Jonge WJM (2005) Unifying ultrafast magnetization dynamics. Phys Rev Lett 95(26):267207. doi:10.1103/PhysRevLett.95. 267207
40. Fahnle M, Illg C (2011) Electron theory of fast and ultrafast dissipative magnetization dynamics. J Phys-Condens Matt 23(49):493201. doi:10.1088/0953-8984/23/49/493201
41. Tserkovnyak Y, Brataas A, Bauer GEW (2002) Spin pumping and magnetization dynamics in metallic multilayers. Phys Rev B 66(22):224403. doi:10.1103/PhysRevB.66.224403
42. Tserkovnyak Y, Brataas A, Bauer GEW, Halperin BI (2005) Nonlocal magnetization dynamics in ferromagnetic heterostructures. Rev Mod Phys 77(4):1375–1421. doi:10.1103/ RevModPhys.77.1375
43. Brataas A, Tserkovnyak Y, Bauer GEW (2011) Magnetization dissipation in ferromagnets from scattering theory. Phys Rev B 84(5):054416. doi:10.1103/PhysRevB.84.054416
44. Mizukami S, Sajitha EP, Watanabe D, Wu F, Miyazaki T, Naganuma H, Oogane M, Ando Y (2010) Gilbert damping in perpendicularly magnetized Pt/Co/Pt films investigated by all-optical pump-probe technique. Appl Phys Lett 96(15):152502. doi:10.1063/1.3396983
45. Tserkovnyak Y, Brataas A, Bauer GEW (2002) Enhanced Gilbert damping in thin ferromagnetic films. Phys Rev Lett 88(11):117601. doi:10.1103/PhysRevLett.88.117601
46. Pal S, Rana B, Hellwig O, Thomson T, Barman A (2011) Tunable magnonic frequency and damping in [Co/Pd]$_8$ multilayers with variable Co layer thickness. Appl Phys Lett 98 (8):082501. doi:10.1063/1.3559222
47. Ganguly A, Azzawi S, Saha S, King JA, Rowan-Robinson RM, Hindmarch AT, Sinha J, Atkinson D, Barman A (2015) Tunable magnetization dynamics in interfacially modified Ni$_{81}$Fe$_{19}$/Pt bilayer thin film microstructures. Sci Rep 5:17596. doi:10.1038/srep17596
48. Mizukami S, Ando Y, Miyazaki T (2001) The study on ferromagnetic resonance linewidth for NM/80NiFe/NM (NM = Cu, Ta, Pd and Pt) films. Jpn J Appl Phys 40(2A):580–585. doi:10. 1143/jjap.40.580

Chapter 4
Experimental Techniques to Investigate Spin Dynamics

As discussed in the Chap. 2, the time scale for magnetization dynamics varies from microseconds (μs) to femtoseconds (fs) which depends on the mechanism involved. In order to understand the magnetization dynamics at various time scales from fundamental perspective and for application in magnetic recording industry, variety of techniques in the frequency, wave-vector, and time domains were invented in last few decades [1, 2]. Historically, conventional ferromagnetic resonance (FMR) [3], which is a frequency domain technique, has been invented first where the sample is excited at a particular frequency and subsequently probed. The external bias field is swept to probe the magnetization dynamics through the ferromagnetic resonance of the sample. Broadband ferromagnetic resonance based on vector network analyzer (VNA-FMR) is also a frequency domain technique but with very high frequency resolution and very good sensitivity over a broadband of frequency [4]. In VNA-FMR, the bias field is kept fixed and the excitation frequency is varied from tens of MHz to tens of GHz range to study the dynamical response of magnetization. A powerful optical technique to investigate the magnetization dynamics is the Brillouin light scattering (BLS) technique where it can be measured in different domains, including the traditional frequency and wave-vector domains [5, 6]. Recent developments of space-resolved, time-resolved, and phase-resolved BLS techniques have taken the application of BLS to very advanced level. Pulse-inductive microwave magnetometry (PIMM) allows time-domain detection of magnetization dynamics with the time resolution limited to few tens of picoseconds [7]. The time-resolved magneto-optical Kerr effect (TR-MOKE) microscopy can attain extremely high time resolution ($\sim$few tens of fs) along with a spatial resolution in the sub-μm regime [8, 9]. The pulse width of the laser limits the time resolution of the TR-MOKE, and till date, it is considered to be most efficient technique to probe ultrafast magnetization dynamics. Furthermore, by incorporating a scanning microscope with the TR-MOKE, time-resolved scanning Kerr microscopy (TR-SKM) [10] has been developed to image the time evolution

© Springer International Publishing AG 2018
A. Barman and J. Sinha, *Spin Dynamics and Damping in Ferromagnetic Thin Films and Nanostructures*, https://doi.org/10.1007/978-3-319-66296-1_4

of spatial distribution of magnetization in confined magnetic elements. In this chapter of the book, we will describe in detail mainly TR-MOKE, BLS, and FMR techniques along with the recent development in the investigation capability of these techniques. All these have wide applicability in the study of spin dynamics. These techniques have enriched the fundamental understanding of magnetization dynamics and provided detailed information on various useful magnetic properties such as saturation magnetization, anisotropy, exchange constant, and most importantly magnetic damping. We describe in this chapter various important techniques used for investigation of spin dynamics in ferromagnetic materials with particular emphasis on thin films and nanostructures. Advantages and disadvantages related to these techniques will also be discussed at the end of this chapter.

4.1 Time-Resolved Magneto-Optical Kerr Effect (TR-MOKE)

As discussed in Chap. 2, a pulsed excitation (magnetic field, thermal, optical, spin torque, etc.) triggers a rich spectrum of spin dynamical processes. Over the years, TR-MOKE technique has emerged as one of the most reliable and acceptable techniques to investigate magnetization dynamics directly in the time domain. In the subsections below, we discuss the basics of TR-MOKE and the experimental approaches, mostly concentrating on all-optical implementation of it [9, 11]. Particular emphasis is given on the development and description of TR-MOKE set up with better sensitivity along with high time resolution in these experiments.

4.1.1 Basics and Background

The magneto-optical Kerr effect (MOKE) discovered by John Kerr in 1877 became an important characterization technique for magnetic thin films and multilayers since its discovery [12]. In particular, MOKE has been used as an efficient tool for measuring the magnetic hysteresis loops and imaging magnetic domain. In 1932, classical theoretical description of MOKE was presented [13]; however, the quantum mechanical picture of this effect was proposed by Argyres in 1955 [14]. In case of interaction of linearly polarized light with a magnetized specimen, its polarization undergoes net rotation which is observed as elliptical polarization in the MOKE experiments. The magneto-optical interaction introduces an orthogonal component k in the electric field vector of the reflected light both in- and out-of-phase to that of the reflected light r. The in-phase component gives rise to the Kerr rotation (θ_k), whereas the out-of-phase component is responsible for the Kerr ellipticity (ε_k). Both of these quantities θ_k and ε_k provide crucial measure of

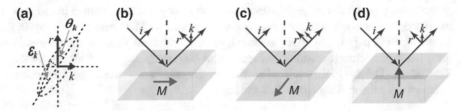

Fig. 4.1 **a** Geometry for the Kerr rotation (θ_k) and Kerr ellipticity (ε_k). **b–d** Different MOKE geometries, **b** longitudinal MOKE, **c** transverse MOKE and **d** polar MOKE. Here, i = incident electric field vector, r = reflected electric field vector, k = induced (due to MOKE effect) orthogonal component in the reflected electric field vector (both in and out of phase to r)

magnetization of the sample (cf. Fig. 4.1a). Depending on the relative orientations of the plane of incidence and magnetization (M) of the sample, there are three important geometries of MOKE as illustrated in Fig. 4.1b–d: (i) longitudinal MOKE, (ii) transverse MOKE, and (iii) polar MOKE. For the longitudinal geometry, M lies in the sample plane and parallel to the plane of incidence. For the polar geometry, M is perpendicular to the sample plane. The longitudinal and polar Kerr effects are characterized by a rotation of the plane of polarization; the amount of rotation is proportional to the component of magnetization parallel to the plane of incidence. The longitudinal and polar MOKE occurs for both p- and s-polarized lights, whereas the transverse effect occurs only for p-polarized light. In the transverse geometry, M lies in the sample plane but perpendicular to the plane of incidence. This effect involves a change in the reflectivity of the light polarized parallel to the plane of incidence, not a rotation of the polarization. This change in reflectivity for the transverse effect depends upon the magnetization component perpendicular to the plane of incidence.

Classically, the physical origin of the MOKE can be explained by considering the magnetic circular dichroism effect [13]. As the light propagates through the medium, its electric field sets electrons into motion. The right circularly polarized (RCP) electric field leads to right-circular electron motion, and left circularly polarized (LCP) electric field drives the electrons in left-circular motion. In the absence of an external magnetic field, the radii of these two circular motions become equal leading to a zero difference in the dielectric constants. However, in the presence of an external magnetic field, the scenario is completely different. The electrons experience an additional Lorentz force due to the external magnetic field. This in turn affects the radii of the left- and right-circular path, and as a consequence, a finite difference appears in the dielectric constants of the left and right circularly polarized modes. The refractive indices of RCP and LCP lights turn out to be different in the presence of the external magnetic field. So, the reflected beam no longer remains a linearly polarized beam, but becomes an elliptically polarized light which is observed as Kerr effect. The angle, by which the major axis of the polarization ellipse is rotated from the original linear polarization axis, is the Kerr angle θ_k.

Following quantum mechanical approach, the Kerr effect is explained in terms of microscopic electronic structure based on the Fermi Golden rule [15] or by the Kubo formalism [14, 16]. It is assumed in this approach that the simultaneous occurrence of exchange splitting and spin–orbit (SO) coupling is responsible for the Kerr effect. The MOKE is related to the off-diagonal components of the optical conductivity tensor. This tensor has the form:

$$\sigma(\omega) = \begin{bmatrix} \sigma_{xx}(\omega) & \sigma_{xy}(\omega) & 0 \\ -\sigma_{xy}(\omega) & \sigma_{xx}(\omega) & 0 \\ 0 & 0 & \sigma_{zz}(\omega) \end{bmatrix} \tag{4.1}$$

Here, the z-axis is perpendicular to the sample. The complex Kerr angle is defined as:

$$\Theta_K \equiv \theta_K + i\varepsilon_K \tag{4.2}$$

For a film of thickness d, its expression is obtained as:

$$\Theta_K = \frac{i\sigma_{xy}}{\sigma_{xx}^s} \frac{4\pi d}{\lambda} \tag{4.3}$$

Here, σ_{xx}^s is the optical conductivity of the substrate, and λ is the wavelength of light. This expression is valid only for $\lambda \gg d$. The real part $\sigma'(\omega)$ and the imaginary part $\sigma''(\omega)$ of the conductivity tensor are even and odd, respectively, with respect to ω and are linked by the Kramers–Kronig relations. The dissipative part of the off-diagonal component of the conductivity tensor (for $\omega > 0$), for an optical transition from the initial state i to the final state f due to the absorption of a photon, is given by:

$$\sigma_{xy}''(\omega) = \frac{\pi e^2}{4\hbar\omega m^2 \Omega} \sum_{if} f(\varepsilon_i) \left[1 - f(\varepsilon_f)\right] \times \left[\langle i|p_-|f^2\rangle - \langle i|p_+|f^2\rangle\right] \delta(\omega_{fi} - \omega) \tag{4.4}$$

where $p_\pm \equiv p_x \pm ip_y$, $f(\varepsilon)$ is the Fermi-Dirac function, Ω is the total volume, and $\hbar\omega_{fi} \equiv \varepsilon_f - \varepsilon_i$. The factor $\delta(\omega_{fi} - \omega)$ takes energy conservation into account. The matrix elements $\langle i|p_-|f\rangle$ and $\langle i|p_+|f\rangle$ correspond to dipolar electric transitions for right and left circularly polarized lights. Clearly, $\sigma_{xy}''(\omega)$ is proportional to the difference of absorption probabilities of the right and left circularly polarized lights and hence gives rise to the Kerr effect.

Similar effect has been observed in transmission as well, for magnetization parallel to the plane of incidence known as Faraday effect, while for perpendicular magnetization it is known as Voigt effect.

The above description of MOKE mainly applies for the static case. For the magnetization dynamics to be investigated by implementing the MOKE (time-resolved magneto-optical Kerr effect, TR-MOKE), the mechanism is quite involved and complicated. In the magnetization dynamics measurement, the initial non-equilibrium in electron distribution acts on dielectric permittivity ϵ. The Kerr angle for the case of magnetization dynamics is given by

$$\theta(t) = F(t)M(t) \tag{4.5}$$

where $F(t)$ denotes the time-dependent Fresnel coefficient which includes all the details of experimental configuration and the sample layout [17, 18]. Thus, it is possible that while measuring $\theta(t)$ in a time-resolved experiment, the perturbation may modify $M(t)$ as well as the generalized Fresnel coefficient $F(t)$. Thus, interpretation of $\theta(t)$ in terms of only $M(t)$ may not be appropriate. Under the assumption that the excitation of the spins is a weak perturbation to the magneto-optic signal, we can write

$$\Delta\theta(t) = F_0\Delta M + M_0\Delta F \tag{4.6}$$

where the Δ represents the laser-induced changes, and the index '0' denotes the state before the perturbation. The ΔF includes all-optical contributions such as state filling. Thus, from the perspective of studying time-resolved magnetization dynamics, optical contribution (ΔF) and the magnetic contribution (ΔM) are intricately related and not straightforward to separate [17]. In 2000, in a theoretical study of the time-resolved magneto-optical Kerr effect, Zhang et al. showed that the temporal difference between the magnetic and the optical response can reach up to about 50 fs [19].

With advances in optics, particularly invention of pico- and femtosecond lasers, tremendous improvement in the measurement of ultrafast dynamics has been witnessed in recent time. One of the main motivations for developing such a technique was to perform FMR experiments in the time domain to verify the LLG equation and to extract the precession frequency and the damping behavior. A more advanced motivation was to directly image the space-time evolution of magnetization. In 1960s, using digital storage oscilloscope, time-resolved magnetization dynamics was studied for the first time where the magnetic field pulse-induced magnetization oscillation with time period in the nanosecond time scale was observed [20, 21]. In 1991, Freeman et al. reported the first picosecond (ps) TR-MOKE measurement of magnetization dynamics of magnetic thin film in which the technique of magneto-optic sampling was used [22]. We briefly discuss the details of the experimental arrangement used in the experiment of Freeman et al. A parallel strip coplanar transmission line structure was fabricated on a semi-insulating semiconductor substrate, and the device was mounted on a chip carrier which was electrically connected to an external bias source. Transient photoconductivity in the semiconductor was created by using optical pulse at the end of the biased transmission line which subsequently launched fast current pulse through it. Due to the opposite signs of the current pulses on the two sides of the lines, a homogeneous magnetic field pulse was created in the gap between the lines which yields a bandwidth of the order of 1THz. By using a probe pulse, which is time delayed with respect to the pump pulse, the transient magneto-optic response of a specimen placed under the transmission lines was measured at the region between the transmission lines. In this experimental arrangement, by scanning the time delay between the pump and probe pulses by an optical delay line, the time-dependent magnetic response of the sample was measured with a time resolution similar to the pulse width of the laser. Note that the rise-time of the pulse is determined by the carrier mobility and the optical pulse width, while the fall-time is governed by the carrier lifetime due to the recombination and sweep-out in the presence of a bias field. Magnetization evolution and relaxation dynamics in pure and Tb-doped EuS thin films were measured using this technique. From this study, it was concluded that the spin–lattice relaxation time becomes an order of magnitude shorter due to the spin–orbit coupling after Tb doping in the EuS film. Freeman et al. in 1992 investigated magnetization dynamics in magnetic insulator YIG film under an in-plane applied bias field [8]. Interestingly, they revealed the precessional spin dynamics in their study as the pulsed magnetic field within the gap between the transmission lines eventually applied perpendicular field which exerted a torque on the spin systems.

4.1.2 All-Optical TR-MOKE Microscope

An important step toward the advancement of the TR-MOKE experiment was to the develop an all-optical TR-MOKE microscope, in which, instead of an electronically or optically generated magnetic field pulse, a fs pump laser itself is used to excite the electron, spin, and lattice systems of a solid and another probe laser is used to follow the ensuing dynamics. In order to realize an all-optical TR-MOKE experiment in a crossed-polarizer configuration, pump and probe beams are focused to overlapping spots on the sample. The pump beam passes through a mechanical delay line to vary the time delay. The influence of the pump beam on the polarization state of the reflected probe beam is measured using an analyzer at an angle (α_A) and an appropriate photodetector. To enhance the measurement sensitivity, a mechanical chopper is usually placed in the path of the pump beam, and a lock-in detection technique is used to measure $\Delta\theta(t)$. In 2003, Koopmans et al. showed that the pump-induced change in the output signal can be described in lowest order in α_A and $\Delta\theta$ [23]. Bigot et al. in 2004 discussed drawbacks of the crossed-polarizer approach and suggested that these may be overcome by performing measurements at a multitude of analyzer angles [24]. Interestingly, based on a two-color collinear pump–probe all-optical TR-MOKE technique, Barman et al. in 2006 measured the ps precessional dynamics of single nanomagnets [11]. In their experiments, the nanomagnets were optically pumped by linearly polarized strong laser pulses $(\lambda \sim 400$ nm, pulse width ~ 100 fs), which causes an ultrafast demagnetization, and subsequently, an internal anisotropy field pulse is created in the system. This results in a change in the equilibrium magnetization orientation and triggers precessional spin dynamics in the nanomagnets. To probe the magnetization dynamics, polar magneto-optical Kerr rotation signal was detected by making use of linearly polarized weak laser beam with $\lambda \sim 800$ nm, which was time delayed with respect to the pump beam. The variation of the optical delay between pump and probe beams and simultaneous measurement of Kerr rotation at each time delay allowed the detection of spin dynamics at each point of the time delay. In their experiments, pump beam was chopped at 2 kHz and they used a phase-sensitive detection using a lock-in amplifier. By careful optical alignment and using a single microscope objective, pump and probe beams were focused down to submicron diameter spots and precisely overlapped at the center of the sample at normal incidence. The optical arrangement allowed the back-reflected pump and probe beams to be collected by a 50:50 beam splitter and made the probe beam to be incident on to the detector. A small part of both pump and probe beams are directed to a CCD camera for viewing the sample position before it can fall on the detector. For viewing the samples, white light illumination is sent to the sample through the same microscope objective. In order to eliminate the pump beam from entering into the detector, a spectral filter is placed before the detector. By introducing an optical bridge detector (balanced photodiode detection), Rana et al. improved the detection system significantly and achieved better spatial sensitivity in all-optical TR-MOKE setups. The optical bridge detector consists of a polarized beam splitter (PBS) and two photodiodes where the difference in the signal between the two photodiodes is proportional to the Kerr rotation. The PBS is placed at an angle of 45° with respect to the reflected light. Thus, in the absence of Kerr rotation, the intensity of linearly polarized light passing through PBS remains identical in two orthogonal components of polarization which results in the so-called balanced condition in the bridge. In the presence of Kerr rotation, the intensities in the two orthogonal components of polarization get modified and give rise to a finite electronic signal at the output of the optical bridge detector. It is important to mention here that for certain materials such as Ni, Kerr ellipticity is much larger than Kerr rotation, and in these cases, a $\lambda/4$ plate is introduced before the analyzer to convert ellipticity into rotation. A schematic of an all-optical TR-MOKE microscope is shown in Fig. 4.2. A typical configuration of the pump and probe beams focused through the same microscope objective and incident on the sample is also shown along with a representative Kerr rotation and reflectivity data. In the latter chapters of this book, the magnetization dynamics data obtained from this setup will be discussed in detail with a particular emphasis on the extraction of magnetization precession frequency and effective damping in various interesting ferromagnetic thin films and nanostructures.

4.1.3 Benchtop TR-MOKE Setup

An important development in making TR-MOKE widely applicable was made by Barman et al. in 2008 when they developed a compact benchtop TR-MOKE setup based on a ps pulsed laser. In this setup, the excitation is performed by a ps pulse generator, and for the detection, an electronically synchronized ps laser pulse is

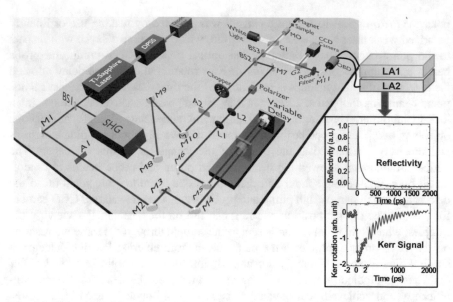

Fig. 4.2 Schematic of the all-optical time-resolved magneto-optical Kerr effect microscope at the S. N. Bose Centre, Kolkata, India. The setup has spatial resolution of ~ 800 nm and temporal resolution of ~ 100 fs, and it is based on two-color (400 and 800 nm) collinear pump–probe technique. The incident pump and probe beams on the sample through the microscope objective are shown at the *right top corner* and the representative reflectivity. The Kerr signal and reflectivity signal obtained for a typical ferromagnetic sample are shown at *right bottom corner*

used. Interestingly, this setup has been custom assembled with a conventional upright microscope configuration with separate illumination, imaging, and magneto-optical probe paths. Important features of this system include high stability, long range of time delay up to 40 ns, and convenient optical alignment. The setup has capability to investigate spin dynamics with 0.5 ps temporal resolution with spatial resolution down to 600 nm. Further details of the benchtop TR-MOKE magnetometer may be found in ref [25]. Using this setup, Barman et al. reported the precessional dynamics of a permalloy microwire and a permalloy microdisk with a vortex state. Interestingly, the gyration mode of a vortex core was measured in the time domain over a sufficiently long time window of about 30 ns, and the corresponding vortex core gyration frequency was found to be 255 MHz. Subsequently, using this technique, the gyration mode splitting of a vortex core due to the magnetostatic interactions from other disks in an array of $Ni_{81}Fe_{19}$ disks of 1 μm diameter, 50 nm thickness, and interdisk separations varying between 150 and 270 nm was demonstrated [26]. They observed the splitting of the vortex core gyration mode when the interdisk (edge-to-edge) separation is 200 nm or less. Also, it was found that a bias magnetic field can control the mode splitting. Through micromagnetic simulation, they interpreted the mode splitting as normal modes of vortex cores analogous to the coupled classical oscillators.

4.1.4 Time-Resolved Scanning Kerr Microscope

Another variant of TR-MOKE, namely time-resolved scanning Kerr microscope (TR-SKM), exploits the combination of ultrafast laser techniques and optical microscopy allowing the observation of repetitive magnetization dynamics. Hiebert et al. in 1997 [10] reported the first TR-SKM study of the non-uniform precessional dynamics in a lithographically patterned $Ni_{81}Fe_{19}$ disk of diameter 8 µm. In their study, the dynamics was excited by pulsed magnetic field created by a biased GaAs-photoconductive switch. A lithographically patterned gold coil was used for sending current pulse around the $Ni_{81}Fe_{19}$ disk which in turn generated the transient magnetic field at the center of the coil along out-of-plane direction. A time-delayed probe beam was used to probe and image the dynamics with a spatial resolution better than 0.7 µm. Interestingly, they observed precessional magnetization dynamics with well-defined damping estimated as 0.008. From the scanning images obtained in the experiment, they analyzed and interpreted the spatially non-uniform dynamics with the initial evolution of Kerr rotation near the edges carrying free magnetic poles, which eventually propagate toward the center. The non-uniform demagnetizing field near the edges of elements carrying free magnetic poles was attributed to be the main reason for non-uniform magnetization dynamics. Stotz and Freeman in 1997 [27] reported a high-resolution scanning optical microscope based upon solid immersion lens for stroboscopic time-resolved studies of magnetic materials. They used hemispherical and the truncated-sphere solid immersion lenses. In an interesting study in 2002, Park et al. [28] using TR-SKM directly observed the localized spin-wave modes in lithographically patterned single $Ni_{81}Fe_{19}$ wires of 2 and 5 µm widths. By compiling successive snapshots of the cross section of these microwires at fixed time steps of 20 or 40 ps, they mapped out full spatiotemporal images of the dynamics in Damon–Eshbach (DE) and backward volume magnetostatic spin-wave (BWVMS) geometries. Interestingly, it was found from frequency domain images that both center and edge modes appear for BWVMS geometry, whereas in the DE geometry edge modes do not appear.

Neudert et al. in 2008 [29] proposed advancement in TR-SKM by introducing phase-locked harmonic excitation of individual resonant modes instead of a broadband excitation by a pulsed magnetic field. In a 40-µm-wide and 160-nm-thick $Fe_{70}Co_8B_{12}Si_{10}$ square element, they identified and imaged two resonant modes by following this technique. In case of excitation with the frequencies of the individual modes, a superposition of both modes with different amplitudes was observed. By detecting out-of-plane and in-plane Kerr rotations, they investigated the relative shape of the magnetization orbit. Although the absolute orbit shape cannot be determined, because of the different dependency of the longitudinal and polar Kerr effect upon the in- and out-of-plane magnetization component, however, from the observation, it was concluded that the relative shape of the magnetization orbit mostly remains unchanged with distance from the edge of the element.

In 2003, Barman et al. performed imaging of magnetization precession in a square $Ni_{81}Fe_{19}$ element with 10×10 μm^2 area and with 150 nm thickness with varying bias magnetic field (**H**) in the plane of the element and exciting the dynamics by a pulsed magnetic field. Interestingly, they observed anisotropy in the apparent damping behavior with varying in-plane orientation of **H**. The damping occurred much faster when bias field is applied parallel to a diagonal of the element as opposed to that when **H** is applied parallel to one of the edges of the element. For applied bias field parallel to the edge and diagonal to the square, the time-resolved Kerr images were thoroughly investigated to get in-depth understanding of damping. It was inferred from the results of this study that anisotropy in the damping is related to spatial non-uniformity of magnetization dynamics at the centre of the element. In case of **H** parallel to the edge of the element, the non-uniformity is confined to the edges until longer delay times. In subsequent studies in 2004 [30] and 2007 [31], Barman et al. through experimental investigation and using simulation showed that spin-wave dephasing is at the origin of larger apparent damping at the center of the element when **H** is applied parallel to a diagonal as compared to when **H** is applied parallel to an edge. Importantly, these studies led to conclusion that the shape of a thin-film element can have a major influence upon the apparent damping of the magnetization precession.

Several other applications of time-resolved scanning Kerr microscope have been reported such as imaging precessional switching, magnetic vortex and domain-wall dynamics, and coherent suppression of small amplitude precession. Interested readers may see the review article [2] for further details.

4.2 Brillouin Light Scattering (BLS)

The light scattering technique has been used for many decades to measure the coherent dynamic properties in gases, liquids, and solids. In this technique, the properties of the scattered light from the sample are characterized and compared to those of the incident light which gives information about the mechanisms that play a role in the scattering process. In the so-called Raman scattering [32], light interacts with rotational or vibrational degree of freedom of the system. The Brillouin light scattering (BLS), named after its inventor, deals with the inelastic scattering of photons from a rather lower frequency (in the order of GHz) excitations, viz phonons, magnons, plasmons. For instance, using this non-contact and thus noninvasive tool, it has been possible to measure the elastic properties in water [33, 34], solids [35], as well as organic materials (like eyes lens) [36].

The theoretical prediction of light scattering from acoustic waves was given independently by Léon Brillouin in 1922 [5] and Mandelstam in 1926 [37]. Subsequently, after few years in 1930, Gross experimentally confirmed the observation of scattering of light in liquids [38]. Later on, the invention of laser in the 1960s revolutionized this area of research; however, the investigation of acoustic waves and spin waves in optically opaque material became possible only after Sandercock in 1971 developed a highly sophisticated spectrometer [39]. The key to this development was noticing the fact of dramatic enhancement of sensitivity of a Fabry–Perot interferometer if the scattered light passes multiple times through it [40]. Both surface and bulk spin waves were detected in polycrystalline films of Fe and Ni by Sandercock and Wettling using tandem operation of two interferometers [41, 42]. In succeeding years, BLS has developed to be a very powerful and versatile tool in magnetic research because of its degree of flexibility in samples, frequency resolution, phase resolution, time resolution, and localized spatial resolution. The other advantages include (i) the ability to measure the thermal excitations (without any external stimulation), that even up to a frequency range as high as 500 GHz, with a resolution of 50 MHz. (ii) the potential to investigate dispersion of spin waves with different absolute values and orientations of their wave vectors. (iii) Further, one can extract rich information about the magnetic properties of magnetic layers, such as saturation magnetization, magnetic anisotropy, and coupling parameter between different magnetic layers.

4.2.1 Principle

BLS is a spectroscopic technique where a beam of highly monochromatic laser light is incident on the surface of the sample under study. The scattering geometry depicting the incident and scattered beams, the incidence angle, and the direction of wave vectors is shown in Fig. 4.3. Although most of the light is specularly reflected or absorbed, a small fraction of the light is scattered from the thermally excited spin waves, which can be divided into two main categories: the elastic scattering and inelastic scattering. In the elastic scattering (like Rayleigh scattering) of photons, the photon's energy or frequency is unchanged. However, in case of inelastic scattering, a shift in the angular frequency takes place, which forms the basis of the spin-wave detection in this technique.

The scattered light is collected (using the same lens as that used for incident beam) within a solid angle in the direction 180° from the incident light, which is

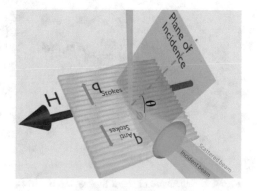

Fig. 4.3 Scattering geometry showing the incident reflected and scattered beams, the direction of magnon wave vectors for Stokes and anti-Stokes process in BLS. The measurement geometry shown is Damon–Eshbach geometry

known as the 180°-backscattering geometry. The backscattered geometry has the advantage that it maximizes the magnitude of the spin-wave wave vector taking part in the scattering process. The light is then frequency-analyzed using a multipass tandem Fabry–Perot (FP) interferometer to extract the information about the surface and bulk spin waves.

From a quantum mechanical viewpoint, the mechanism of inelastic scattering can be stated as a photon–magnon collision, i.e. in terms of the creation (Stokes process) and annihilation (anti-Stokes process) of a magnon of wave vector q and angular frequency ω. The process is shown in Fig. 4.4. Considering the conservation of energy (frequency) and momentum (wave vector) between the magnon and the incident (i) and scattered (s) photons, it follows

$$\hbar\omega_s = \hbar\omega_i \pm \hbar\omega \qquad (4.7)$$

$$\hbar k_s = \hbar k_i \pm \hbar q \qquad (4.8)$$

where '+'('−') sign stands for the anti-Stokes (Stokes) shift, and ω_i, k_i, ω_s, k_s are the respective angular frequencies and wave vectors of the incident and scattered light. Note that for light scattering from thin films, the perpendicular component of the wave is not conserved because of translational symmetry breaking. Hence, the above equations are only valid for q which is the wave-vector component parallel to the film plane. From the conservation of momentum described above, it is possible to determine the wave vector of magnon taking part in the scattering process. Basically, the amount of energy of incident light exchanged with the system during the scattering (i.e. the energy of the magnon itself) is very small with respect to the incident photon energy. The energy of a visible photon is few eV, whereas the energy of magnon observed in BLS is about $\sim 10^{-4}$ eV. Consequently, the magnitude of the wave vector of the scattered photon (k_s) should be very close to that of the wave vector k_i of the incident photon. Figure 4.5a schematically illustrates the scattering profile of a photon by a bulk magnon, which means that the magnon taking part in the scattering has a component perpendicular to the surface. Here, the scattered photon wave vector k_s must lie on the dashed circle line, whose radius is equal to the magnitude of incident photon wave vector (k_i). Here, the cone represents the collection angle of the scattered beam in the BLS experiment, whose central axis aligns

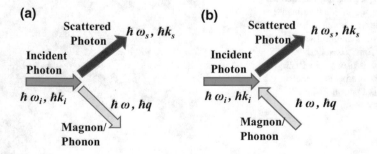

Fig. 4.4 Schematic of the **a** Stokes and **b** anti-Stokes scattering process occurring in BLS

Fig. 4.5 Scattering of laser beam by **a** bulk magnon and **b** surface magnon. The direction of q corresponds to the anti-Stokes process

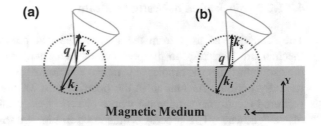

Magnetic Medium

with the incident photon wave vector. Therefore, by assuming that k_i and k_s are collinear ($k_s = -k_i$), i.e. the backscattered geometry, the wave-vector magnitude of the emitted or absorbed bulk magnon $|q|$ is always equal to $2|k_i|$.

On the other hand, the scattering of a photon by a surface magnon is illustrated in Fig. 4.5b. Here, since the direction of q lies in the horizontal plane, the momentum will be conserved only in the plane along the sample surface. In other words, the conserved component of the incident beam is equal to $|k_i| \sin \theta$, where θ is the angle between k_i and sample surface normal (see Fig. 4.3). Therefore, the wave vector of the magnon probed by the experiment is given by

$$|\overline{q}| = |\overline{k_i} \sin \theta - \overline{k_s} \sin \theta| = 2|k_i| \sin \theta \qquad (4.9)$$

The above discussion suggests that by varying the incident angle, no new information can be obtained about the bulk magnon, whereas the surface magnon reveals the important frequency–wave vector dispersion relation. Equation (4.9) is known as the Bragg's condition and provides an approximation for most of the light scattering experiments. According to Eq. 4.9, one can vary the magnitude of q by varying either λ or θ.

4.2.1.1 Uncertainty in the Selected Spin-Wave Wave Vector

Due to the finite aperture angle of the objective lens used for focusing and collecting light from the sample, an uncertainty is always induced in the selected spin-wave wave vector. For finite angle of incidence θ, the corresponding spread in q is given by

$$\Delta q = 2k_i \cos \theta \sin \left(\frac{\phi}{2}\right) = 2\frac{2\pi}{\lambda} \cos \theta NA \qquad (4.10)$$

where φ is the collecting angle of the lens, and NA = sin $(\varphi/2)$ is its numerical aperture (see Fig. 4.3). Clearly, this uncertainty is maximum for normal incidence ($\theta = 0$). Noticeably, a possible uncertainty can also be present in θ due to the focusing of incident beam. However, to reduce this in the practical experiment, a very narrow beam (width ~ 500 μm) is used, which in turn increases the focused spot size.

4.2.1.2 Polarization of Scattered Beam

The scattering of light from the spin waves can be viewed as a magneto-optic mechanism. Basically, in the presence of precessing magnetization or spin wave, the oscillating electric dipoles experience a Lorentz force which effectively causes a spatially periodic fluctuation in the polarizability of the medium. This leads to a scattered electromagnetic wave whose electric field vector is perpendicular to that of incident wave. This process is illustrated in Fig. 4.6 for an incident light which is p-polarized (polarization parallel to the plane of incidence). The geometry shown is the Damon–Eshbach (DE) geometry where the propagation of spin wave is perpendicular to the magnetization direction. As the laser beam hits the sample, it provokes the electric dipoles to oscillate due to its oscillatory electric field, given by $E = E_x e_x + E_y e_y$. Now, say, magnetization M also contains a dynamic component, then M can be written as $M = M_0 + m$, where $m = m_x e_x + m_y e_y$. The fluctuating component of M exerts a Lorentz force (proportional to $E \times m$) on the electric dipoles, which in this case results in the polarization pointing in the z direction. As a result, the radiated light wave has its electric field along the z direction, i.e. s-polarized. A similar argument for s-polarized incident light follows that the scattered light is p-polarized. Therefore, we conclude that the polarization of light scattered by a magnon is perpendicular to the polarization of the incident light. This is different from the case when light is scattered by acoustic phonons, there the polarizations of incident and scattered beam lie in same direction. This helps to isolate the light scattered from magnons and that scattered from phonons by selecting proper orientation of the analyzing polarizer in the experiment.

4.2.2 Experimental Setup

As illustrated in the previous section, the inelastically scattered beam carries information about the frequency and wave vector of the involved spin-wave mode.

Fig. 4.6 Schematic of the interaction between p-polarized incident beam and the precessing magnetization

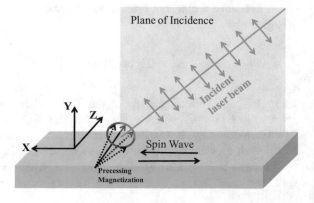

Further, the intensity of the scattered light is proportional to the intensity of studied spin wave. In this section, we will explain how these quantities are accessed and analyzed experimentally. In general, the BLS technique offers two different measurement geometries: (i) forward scattering geometry, where the scattered beam is collected after transmission of the probing beam through a transparent sample, and (ii) back scattering geometry, where the beams that are backscattered from the surface of opaque sample are investigated. The latter can be further extended to BLS microscopy (micro-BLS), where the profile of spin wave can be mapped in a space-, time- and phase-resolved manner. In the following, we will discuss the details of conventional backscattered geometry, followed by the distinguished features of the micro-BLS technique.

Figure 4.7 presents the schematics of the optical beam path used for conventional BLS experimental arrangement. The main components constituting the BLS apparatus include a tandem Fabry–Perot interferometer (TFPI). For the measurement of thermally excited magnons, a laser light emitted by a diode-pumped, frequency-doubled, single-mode solid-state laser of wavelength of $\lambda = 532$ nm is used (the power of the emitted light is 300 mW). Right in front of the laser, the light is split into two beams using a 10:90 beam splitter (BS). The smaller part of the deflected beam is directed straight to the TFPI using mirrors M1 and M2, where it serves as a reference beam. The purpose of this reference beam is manifold:

(i) Firstly, this beam is used to stabilize the mirror spacing of the Fabry–Perot etalons.

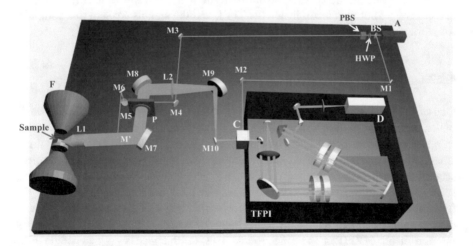

Fig. 4.7 Setup and optical pathway for conventional BLS setup. The notation for the components is given in the text. *M* mirrors, *L* lenses, *P* polarizer, *PBS* polarized beam splitter, *HWP* half wave plate, *F* magnet

(ii) This beam forms the central elastic peak in a BLS spectrum, where it is used
      to determine the frequency shift of the scattered beam *w.r.t.* the incident
      beam.
(iii) Also, reference beam is employed to estimate the transmission order of the
      Fabry–Perot etalon which in turn helps to deduce the frequencies present in
      the scattered light.

The other part of the beam is first sent through a HWP and a PBS, in order to
eliminate the small in-plane polarized component from the partially polarized beam
of the laser. The beam perpendicularly polarized to the optical table is then guided
by the mirrors M3, M4, M5, M6, and M′ toward the sample which is lying between
the electromagnets. The mirror M′ is taken as a tiny prism mirror so as to minimize
the blocking of the backscattered beam by itself. Finally, the beam is focused onto
the sample using an achromatic doublet lens. The sample is mounted on a rotation
stage where change in the rotation angle changes the angle of incidence, thereby
addressing different transferred spin-wave wave vector according to Eq. 4.9.
A magnetic field is applied perpendicular to the transferred wave-vector direction,
i.e. in the DE geometry. The measurements are performed for various magnetic
field values at different wave vectors of transferred spin wave. Subsequently, the
scattered beam is collected by the same lens which is then targeted toward the
entrance unit of a JRS Scientific Instrument (3 + 3)-pass tandem FPI for frequency
analysis, using the focusing lens L2. In order to select the light scattered from spin
waves, a crossed polarizer P is inserted in the path of the scattered light. This allows
for the suppression of the elastically scattered beam as well as the beam containing
the signal from phonons. The resulting light has very low intensity which is
detected by a photon detector inside the TFPI, and the signals are sent to a computer
for storage and analysis.

4.2.2.1   Tandem Fabry–Pérot Interferometer (TFPI)

One of the key issues of BLS spectroscopy is the frequency analysis of spin wave,
which requires very high spectral resolution. Basically, the observed frequency for
spin wave is, at most, 300 GHz, which is about 10 cm^{-1}. This is nearly 10^{-5} times
smaller than that of a typical excitation frequency of laser light. Moreover, the cross
section of the inelastic scattering of photons is very small as compared to the elastic
scattering. Therefore, a high contrast is required for an efficient detection of the
fractional amount of incident laser power with high signal-to-noise ratio. These
conditions are fulfilled by the implementation of a triple-pass tandem Fabry–Pérot
interferometer (TPFI) in the BLS setup. The TFPI consists of two single FPIs
connected in series, and the light passes through each FPI three times, as shown in
Fig. 4.8. In the following, first we address briefly the transmission characteristics of
a single FPI, and then, the realization and the operations of the tandem mode are
discussed.

Fig. 4.8 Schematic of the optical arrangement of TFPI

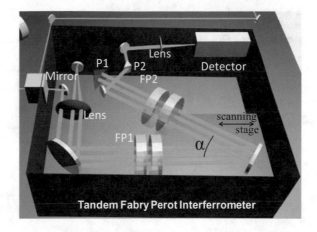

4.2.2.2 The Fabry–Pérot Interferometer

A typical FPI [43] or etalon is constituted of two planar, partially reflecting mirrors mounted accurately parallel to each other at a distance (L). The light entering the FPI undergoes multiple back-and-forth reflections and transmissions. The transmitted beams interfere with each other and results in the condition for constructive interference under normal incidence given by:

$$L = \frac{n\lambda_0}{2} \tag{4.11}$$

where $n = 1, 2, 3\ldots$ is an integer (transmission order), and λ is the wavelength of the light. Therefore, the consecutive orders of interference are separated by a frequency gap Δf as

$$\Delta f = \frac{c}{2L} = \frac{150}{L} \text{GHz.mm}^{-1} \tag{4.12}$$

Here, c denotes the velocity of light. This interorder spacing is known as the free spectral range (FSR) of the interferometer. The finesse of the interferometer is related to FSR as

$$F = \frac{\Delta f_{FSR}}{\Delta f_{FWHM}} \tag{4.13}$$

where Δf_{FWHM} is full width at half maximum of the transmission curve (Fig. 4.9). The finesse F affects the transmitted intensity (I_t) off the FPI via:

Fig. 4.9 Periodic
transmission spectrum
depicting the FSR

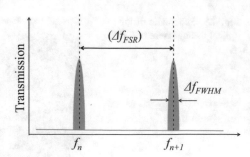

$$I_t = \frac{I_0}{1 + (4F^2/\pi^2)\sin^2(2\pi L/\lambda_0)} \tag{4.14}$$

where I_0 is the intensity of the incident light.

The above equation is known as Airy function which describes the periodicity of transmitted intensity with mirror spacing and frequency. The finesse can be regarded as a measure for the quality of the instrument, which is connected to the reflectivity R of the FP etalons, by the relation

$$F = \frac{\pi\sqrt{R}}{(1-R)} \tag{4.15}$$

Therefore, a higher reflectivity enhances the finesse and as obtained from Eq. 4.15 increases the frequency resolution, as FWHM decreases. On the other hand, for fixed R, an increase in L reduces the FSR, which improves the frequency resolution as the finesse remains constant. The contrast of a FPI is defined as

$$C = 1 + \frac{4R}{(1-R)^2} \tag{4.16}$$

The contrast for an n-pass interferometer is the nth power of that of a single-pass one. For example, a five-pass interferometer can achieve a contrast of at least five orders of magnitude greater than that of a single-pass interferometer.

4.2.2.3 Tandem Operation

One issue inherent to FPI is the periodicity of transmitted intensity as a function of the mirror spacing. The fact that the transmission characteristic repeats every FSR creates certain problems in the identification of the frequencies present in the scattered light. For example, say, for a fixed mirror spacing L, there are two wavelengths present in the measured light beam, such that

$$2L = m_1 \lambda_1 \tag{4.17}$$

$$2L = m_2 \lambda_2 \tag{4.18}$$

Therefore, the transmission condition is satisfied for both the wavelengths at different orders. Since the order of transmission spectrum is determined from the reference beam, the order of second wavelength (which does not match with that of the reference beam) and consequently the wavelength itself remains unaccessed. Moreover, it is difficult to unambiguously identify whether a peak signal belongs to the Stokes side of a specific transmission order or it is the anti-Stokes signal of the previous order. To address these shortcomings, the interferometer is used in a tandem configuration, wherein the light passes consecutively through two interferometers, which are mounted under an angle α. This arrangement was developed by Sandercock and is shown in Fig. 4.10. The right mirror of each FPI sits on the translation stages and the other on a separate angular orientation device. The scanning stage can move the right mirror of each pair along the optical axis of FP1. A displacement d of the translation stage leads to a change of the mirror distance in FP1 by $\Delta L1 = d$, while the change for FP2 is given by:

$$\Delta L2 = \Delta L_1 \cos(\alpha) \tag{4.19}$$

This also satisfies the synchronization condition as

$$\frac{\delta L_1}{\delta L_2} = \frac{L_1}{L_2} \tag{4.20}$$

Now, this can be used to suppress the intermixing of different orders. To do that, before the scanning of linear stage, each FPI is adjusted to transmission separately. Although their transmission orders are different, they together provide a central transmission order, which can be adjusted by changing the mirror spacing of FP2

Fig. 4.10 Transmission spectra of FP1 and FP2 in tandem operation

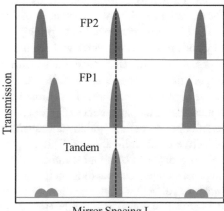

(see Fig. 4.10). At the same time, the other orders are suppressed, since the FSR of the two FPIs is now different. This arrangement also increases the FSR of the final spectrum, without affecting the resolution.

Now, when the stage is moved, the mirror spacing changes by

$$l_1 = (L_{10} + d) \tag{4.21}$$

$$l_2 = (L_{10} + d) \cos \alpha \tag{4.22}$$

where L_{10} is the initial mirror spacing of FP1, i.e. at $d = 0$. This removes the aforementioned ambiguities since the transmission order is now fixed. As mentioned earlier, the light additionally passes through each FPI three times to enhance the contrast. Finally, after the six passes through the FPIs, the light is directed to a photomultiplier, which counts the number of transmitted photons as a function of the mirror spacing and, consequently, as a function of the frequency shift. To achieve this, the scanning stage constantly sweeps the distance which corresponds to the required frequency window and the data is recorded for long time to attain sufficient statistics. In this way, the obtained BLS intensity is proportional to the spin-wave intensity at a given frequency.

4.2.3 BLS Microscopy

4.2.3.1 Micro-BLS—Space-Resolved BLS

The micro-BLS is an extension of conventional BLS which was developed in the past decade [44, 45]. The conventional BLS is an efficient technique for examining spin wave in thin films and large arrays of microstructures but is limited when single micrometer-sized elements are introduced. This is because the spot size of the focused laser is as large as tens of micrometers. In micro-BLS, the laser is focused down to a diffraction limited spot ($\sim$250 nm) on the sample, which allows for a spatial sensitivity, however, at the cost of wave-vector resolution. The reason is the Heisenberg's uncertainty principle, which prohibits the simultaneous access to position and wave vector with arbitrary high precision. The components needed to transform the BLS to a micro-BLS are shown in Fig. 4.11. The additional important components are as follows: a microscope objective, a CMOS camera, a set of high precision three-dimensional translation stages, and a number of steering optics. Here, the laser beam is broadened using two lenses of small and large focal lengths in succession, which subsequently fills the back aperture of the microscope objective of magnification of 100×, a numerical aperture (NA) of 0.75, and a long working distance of about 4 mm. Finally, the beam hits the sample in a range of angle of incidence depending on the numerical aperture $NA = n \sin\theta$, where n is the index of refraction of the medium (1.00 for air), and θ is the angle of incidence. In the setup developed at the S N Bose National Centre for Basic Sciences Kolkata,

Fig. 4.11 Schematic of the optical setup of micro-BLS required for the transformation from conventional BLS

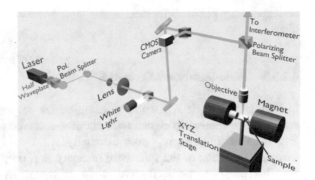

NA = 0.75 and the maximum angle of incidence is given by 48.6°. This corresponds to a range of transferred wave vector from 0 to 1.77×10^5 rad/cm. The scattered beam is collected by the same objective lens (backscattered geometry) and sent to the TFPI for frequency analysis. The purpose of the CMOS camera is twofold: (i) It helps to locate the laser spot on the sample, and (ii) it allows for the stabilization of the sample position with respect to the thermal drift.

In order to conduct a fine spatial scan and for thermal stabilization, the sample is mounted on a set of very high precision translation stages. The spatial resolution of lateral and vertical stages is 1 and 50 nm, respectively. By moving the position of the laser onto the sample, the spatially dependent amplitude of scattered light is measured for a given frequency. To enhance the signal-to-noise ratio, the density of measured magnons is increased by electrical excitation with a RF sent through a coplanar waveguide (CPW).

4.2.3.2 Phase-Resolved Micro-BLS

As pointed out above, the spatial resolution in micro-BLS comes with an uncertainty in the measurement of the wave vector of the spin wave. This limitation can be countered by analyzing the spin-wave phase [46–48]. Probing the propagating spin-wave phase is basically equivalent to the measurement of wavelength and thus the wave vector of spin wave. The information about the spin-wave phase is extracted by analyzing the phase of the scattered beam. This is realized via the interference of the scattered beam with a temporally coherent reference beam that is generated using an electro-optical modulator (EOM). By changing the probing position across the sample, the propagation of the spin-wave phase can be monitored. The intensity of the corresponding interference signal can be expressed as

$$I_{\text{int}} \propto E_S^2 + 2E_R E_S \cos(\Delta\varphi_0 + q.r) + E_R^2 \qquad (4.23)$$

where E_S and E_R are the electric fields corresponding to the sample and the reference beam, respectively, $\Delta\varphi_0$ is the initial phase difference between the reference beam and the sample beam. The argument of the cosine function indicates the corresponding phase difference as the spin wave of wave vector q advances a distance r.

4.2.3.3 Time-Resolved Micro-BLS

In BLS technique, it is possible to examine the propagation and relaxation characteristics of spin-wave eigenmodes via time-resolved analysis [49–51]. The main idea is to excite the spin waves externally at a time t_0 and then to measure the temporal evolution of the spin-wave intensity as it propagates through the sample. For this purpose, the spin-wave mode is repeatedly excited using a train of microwave pulses. The leading edge of the pulse defines the starting time t_0, and the data is acquired at different points of times t_i w.r.t. t_0. The corresponding counts for subsequent pulses are added up which finally renders the matrix formed by the intensity of total accumulated photons and the associated time t_i. Further, one can acquire a two-dimensional intensity map of a spin-wave pulse excited at a fixed frequency for different delays, which basically presents the spatial and temporal advancement of the corresponding wave packet.

It is worth to mention here that a number of research groups have implemented BLS technique in backscattered geometry and BLS microscopy for investigating spin waves and its dispersion in various ferromagnetic thin films and arrays of nanostructures. In particular, using a custom-built BLS setup at S N Bose Centre in India, spin waves in exchange spring ferromagnetic bilayer thin films [52–54], magnetic damping in Heusler alloy thin films [55], magnetic inhomogeneity [56], and asymmetric spin-wave propagation [57] in thin-film heterostructures have been recently demonstrated.

4.3   Ferromagnetic Resonance (FMR)

One of the oldest and widely used techniques to investigate magnetization dynamics is the ferromagnetic resonance (FMR) spectroscopy. Among various experimental techniques developed for investigating magnetization dynamics, FMR is quite well understood. The collective dynamics of the spins in a ferromagnetic material in an external magnetic field describes the FMR [3]. An analytical description can thus be obtained by treating the individual spins as a macroscopic spin. It is interesting to note that the FMR condition describes a spin wave with an infinitely large wavelength, which corresponds to wave vector $q = 0$. Thus, in spintronic and magnonic devices, the FMR frequency represents an important distinction between propagating and localized oscillation modes. Below, we discuss the basic principle of the FMR and the experimental implementation of this technique. We refer other existing literatures to readers for the extensive mathematical formalism of FMR.

4.3.1   Basics

In the FMR experiments, an rf-magnetic field is applied to a ferromagnetic (FM) material under a steady bias field in such a way that the rf-field is perpendicular to the bias field. The resonance phenomenon takes place when the angular frequency of the rf-field becomes equal to the frequency of precession of magnetization in the FM. The magnetization in the FM material precesses with the resonance frequency by absorbing power from the rf-field. Experimentally, in 1946, Griffiths first observed the FMR in a measurement analogous to the Purcell–Torrey–Pound nuclear resonance experiment [58]. Surprisingly, in their study, it was found that the resonance frequency (ω_0) is quite higher than the expected Larmor frequency (ω_L), as given by the following equation, under the same effective field (H_{eff}):

$$\omega_L = \gamma H_{eff} \tag{4.24}$$

The explanation of the anomaly was given by Kittel [3] where it was proposed that it is important to consider the dynamical coupling caused by the demagnetizing field. Kittel [3] performed the calculations by assuming that the magnetization is uniform throughout the sample which is well known as the macrospin formalism. Under this assumption, the magnetic moments of the entire sample can be replaced by single giant macrospin. For any FM material with magnetization M under a bias field H_{eff}, the equation of motion is given by Eq. 4.25:

$$\frac{dM}{dt} = -\gamma(M \times H_{eff}) \tag{4.25}$$

For a general ellipsoid as shown in Fig. 4.12, let us assume that the demagnetizing factors along the three principal axes X, Y, and Z are N_x, N_y, and N_z, respectively. If the bias field is along Z-axis (H_z), and the rf-field is along X-axis (H_x), then the effective values of the magnetic field components are given by:

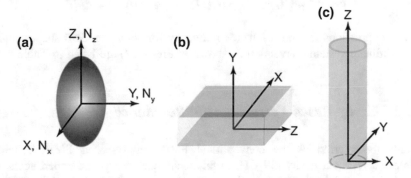

Fig. 4.12 Coordinate systems used for **a** an ellipsoid, **b** a plane, and **c** a cylinder

Table 4.1 Resonance frequencies and demagnetization factors for some standard shapes

Shape	Magnetization direction	Demagnetizing factors			Eigen frequencies
		N_x	N_y	N_z	
Infinitely thin plane	Tangential	0	4π	0	$\omega_0 = \gamma[H_z(H_z + 4\pi M_z)]^{1/2}$
	Normal	0	0	4π	$\omega_0 = \gamma[H_z - 4\pi M_z]$
Infinitely thin cylinder	Longitudinal	2π	2π	0	$\omega_0 = \gamma(H_z + 2\pi M_z)$
	Transverse	2π	0	2π	$\omega_0 = \gamma[H_z(H_z - 2\pi M_z)]^{1/2}$
Sphere	–	$4\pi/3$	$4\pi/3$	$4\pi/3$	$\omega_0 = \gamma H_z$

$$H_x^i = H_x - N_x M_x \tag{4.26}$$

$$H_y^i = -N_y M_y \tag{4.27}$$

$$H_z^i = H_z - N_Z M_Z \tag{4.28}$$

Substituting Eqs. 4.26–4.28 into Eq. 4.25:

$$\frac{dM_x}{dt} = \gamma \left[H_z + \left(N_y - N_z \right) M_z \right] M_y \tag{4.29}$$

$$\frac{dM_y}{dt} = \gamma [M_z H_x - (N_x - N_z) M_x M_z - M_x H_z] \tag{4.30}$$

$$\frac{dM_z}{dt} \cong 0 \tag{4.31}$$

By considering time-dependent variation of M and H ($\exp(j\omega t)$), the equation of resonance frequency is given by:

$$\omega_0 = \gamma \sqrt{\left[H_z + \left(N_y - N_z \right) M_z \right] \times \left[H_z + (N_x - N_z) M_z \right]} \tag{4.32}$$

The expressions for the resonance frequency for some standard shapes (under the coordinate system convention as shown in Fig. 4.12) are listed in Table 4.1.

4.3.2   Cavity-Based Ferromagnetic Resonance

Commonly referred as the conventional FMR, cavity-based FMR technique employs a resonant cavity [59–61]. In this technique, during the experiments, the sample is placed in a resonant microwave cavity. A Klystron or a Gunn diode provides microwave pumping field with a fixed frequency (typically $1 \sim 80$ GHz), which is coupled via a waveguide into the cavity creating a standing microwave field. The sample is placed into a maximum of the magnetic field component of the standing microwave field, acting as the pumping field. Different pumping geometries may be realized for various sample positions depending upon the type of the cavity used. Typically, an electromagnet is used to generate an external magnetic field. The cavity is placed between the pole pieces of an electromagnet which applies the bias field. In a typical experiment, the microwave frequency is kept constant, while the applied bias field is swept. When the resonance condition is satisfied, then the power gets absorbed by the sample from the microwave radiation. The reflected radiation from the cavity measured using a microwave diode shows the resonance signature. The absorbed power is a measure of the imaginary part of the complex susceptibility χ, or alternately, the measured signal is proportional to the out-of-phase microwave susceptibility χ'' where $(\chi = \chi' - i\chi'')$. Following detailed mathematical calculations, the out-of-phase microwave susceptibility has been found to be typically of resonance Lorentzian shape. The maximum absorption occurs at the FMR field (H_{res}), and the linewidth of the resonance is defined by microwave losses. During the experiments, to enhance the measurement sensitivity, the bias field is modulated with an alternating modulation field (typically $100 \sim 200$ Hz). Care is taken to ensure that the modulation field is appreciably smaller than the FMR linewidth. This yields the absorption derivative spectrum and subsequently allows the measurement of the signal using lock-in amplifier-based detection. The measured signal is proportional to the field derivative $d\chi''/dH$. The resonance field corresponds to the zero crossing of the $d\chi''/dH$, and the FMR linewidth is given by the field interval between the two extrema of $d\chi''/dH$. It is important to mention here that the FMR technique has been widely used for accurate determination of magnetic anisotropies, g-factors, and magnetic damping parameters of ultrathin magnetic structures. Particularly, the linewidth responses are often interpreted in terms of a combined inhomogeneous line broadening and Gilbert damping model. Thus, a field swept half power linewidth has the following form

$$\Delta H = \Delta H_0 + \frac{4\pi\alpha f}{|\gamma|} \tag{4.33}$$

where ΔH_0 represents the inhomogeneous broadening in a field which affects the FMR response, and α is the Gilbert damping parameter [62]. Moreover, in 1991, Celinski et al. [62] demonstrated the ability of FMR technique to determine the layer-averaged magnetic moment of an ultrathin film by comparing FMR data collected for the film with that of a reference specimen of known magnetization.

4.3.3   Broadband Ferromagnetic Resonance

In 2003, Denysenkov and Grishin developed a continuous wave broadband ferromagnetic resonance spectrometer based on an X-band microwave reflection cavity
and a vector network analyzer (VNA) [4]. The broadband FMR (popularly known
as VNA-FMR) technique allows for operation over a wide frequency band and
yields FMR parameters from standard microwave S-parameter measurements versus frequency and field. Briefly, in waveguides supporting non-TEM propagation,
voltages and currents cannot be specified uniquely; however, they are defined in
terms of scattering parameters, the so-called S-parameters. The S-parameters relate
to the incident and reflected electromagnetic waves. In VNA-FMR experiments, the
microwave frequency is varied at a fixed static field and the FM absorption profile
from standard S-parameter measurements is extracted. Finally, in these experiments, by analyzing the full width at half maximum, the frequency swept linewidth
is obtained, which is further analyzed to extract the characteristics of the sample.
We discuss the experimental setup and results of Kalarickal et al. [60] in which a
50-nm-thick $Ni_{80}Fe_{20}$ film with 5-nm Ta seed layer deposited on a glass substrate is
investigated. The typical experimental arrangement for performing the VNA-FMR
study on this sample is reproduced in Fig. 4.13. The microwave drive in the
experiment is provided by a CPW excitation structure, and the thin-film sample is
positioned across the central conductor. The static magnetic field is provided by the
set of Helmholtz coils. In the above experiment, the CPW had a 100-μm-wide
central line and the static field was applied in the plane of the film and perpendicular
to the microwave field. The setup was then used to obtain the standard microwave
S-parameters as a function of the microwave frequency at a fixed bias magnetic
field for the CPW line with the sample placed in appropriate position. Input and
output signals to and from the CPW were sent and received using VNA, and
subsequently, S_{21} parameter is measured in the transmission geometry. In a

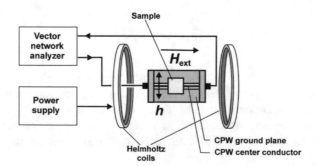

Fig. 4.13 Schematic diagram of the VNA-FMR spectrometer. The sample is placed on the
coplanar waveguide (CPW) structure, as indicated. The mutually perpendicular static applied field
H_{ext} and the microwave field h are in the plane of the film. *Reprinted with permission from
Ref. [60]. Copyright 2006 by the American Institute of Physics*

two-terminal VNA, S_{21} is a complex number defined as the ratio of the received voltage in port 2 and the input voltage in port 1.

In Fig. 4.14, we show the raw experimental data obtained for 50-nm-thick $Ni_{80}Fe_{20}$ film in the frequency range of 1–3 GHz using the VNA-FMR setup. Here, we briefly discuss the analysis procedure of these data as it is slightly involved. The raw data obtained in the VNA-FMR measurement is analyzed based on a transmission line model developed by Barry under the assumption that the dominant CPW mode was the TEM mode. In 1986, Barry [63] proposed analysis (under the assumption that reflection is negligible) which provides an uncalibrated effective microwave permeability parameter of the following form:

$$U(f) = \pm \frac{i \ln[S_{21-H}(f)/S_{21-\text{ref}}(f)]}{\ln[S_{21-\text{ref}}(f)]} \qquad (4.34)$$

where f corresponds to common frequency value at which S_{21} parameters for FMR measurement ($S_{21-H}(f)$) and reference measurement ($S_{21-\text{ref}}(f)$) are performed. Note that the reference data measurement is required to obtain the exact response properties of the excitation structure, feed cables, etc. Only after subtracting the

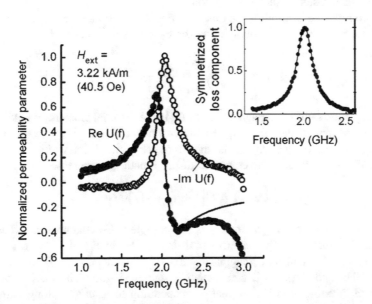

Fig. 4.14 Representative vector network analyzer ferromagnetic resonance (VNA-FMR) data that show the normalized permeability parameter U vs frequency f for film 50 nm $Ni_{80}Fe_{20}$ thin film at an static applied field H_{ext} = 3.22 kA/m (40.5 Oe). The solid circles show the Re[$U(f)$], and the open circles show the -Im[$U(f)$] values extracted from the experimental S parameters. The solid curves show fits to the data. The inset shows the data in a normalized loss component format, with conversion based on the same fit parameters used to obtain the solid curves in the main figure plot. The solid curve in the inset shows the theoretical loss profile. *Reprinted with permission from Ref. [60]. Copyright 2006 by the American Institute of Physics*

reference data, the appropriate FMR response for a ferromagnetic sample may be obtained.

Ideally, in the VNA-FMR measurement, the variation of $-\mathrm{Im}[U(f)]$ versus f corresponds to the FMR loss profile, whereas $\mathrm{Re}[U(f)]$ corresponds to dispersion. Note that the raw measured response is related to the actual complex microwave permeability in these experiments, and conventionally, the choice of sign is followed to keep $\mathrm{Im}[U(f)]$ value negative in the vicinity of FMR peak. In Fig. 4.14, the open and solid circles represent the data for $\mathrm{Im}[U(f)]$ and $\mathrm{Re}[U(f)]$ as reproduced for Kalarickal et al. [60]. These are normalized to obtain a maximum $\mathrm{Im}[U(f)]$ value of unity at the FMR peak. It is interesting to note that the responses observed in their experiments for $\mathrm{Im}[U(f)]$ and $\mathrm{Re}[U(f)]$ do not correspond strictly to the loss and dispersion profiles as predicted by FMR theory. Also, they found asymmetric $\mathrm{Im}[U(f)]$ response, which drops below zero at low frequencies and a significant departure from a dispersive response above about 2.3 GHz for the case of $\mathrm{Re}[U(f)]$. These behaviors in their experiments were mainly due to a small difference between the FMR field value and reference field value and also due to the negligence of reflections in the simplified analysis that gives Eq. 4.35. A mixing of the real and imaginary components of the susceptibility $\chi(f)$ in the measurements and the combination of offsets and distortions due to the FMR response embedded in the reference data result in such observations.

The complex susceptibility response at a frequency f for a uniaxial ferromagnetic thin film magnetized to saturation along the easy axis by a static external field H_{ext}

$$\chi(f) = \left(\frac{|\gamma|\mu_0}{2\pi}\right)^2 \frac{M_S(H_{\mathrm{ext}} + H_k + M_S)}{[f_{\mathrm{res}}^2 - f(f - \Delta f_{\mathrm{VNA}})]} \tag{4.35}$$

where M_s is the saturation magnetization, H_k is the uniaxial anisotropy field, H_{ext} is the external applied bias field, f_{res} is the resonance frequency, γ is the electron gyromagnetic ratio, and Δf_{VNA} is the frequency swept linewidth obtained in VNA-FMR experiments. The complete fitting function is expressed as

$$U_{\mathrm{fit}}(f) = C\left[1 + \chi_0 + \chi(f)e^{i\phi}\right] \tag{4.36}$$

where C is the real scaling parameter, χ_0 is a complex offset parameter, and ϕ is the phase shift adjustment. The experimental data points in Fig. 4.14 were fit simultaneously to both real and imaginary parts of the function $U_{\mathrm{fit}}(f)$ to obtain the $f_{\mathrm{res}} = 2$ GHz and $\Delta f_{\mathrm{VNA}} = 236 \pm 12$ MHz values. Figure 4.14 inset shows the same data in normalized loss format corresponding to $-\mathrm{Im}[\chi(f)]$ along with theoretical fit (shown using solid curve) where the Lorentzian shape indicating microwave loss can be noticed.

Note that the linewidth obtained in the VNA-FMR measurements is frequency linewidth which may not directly correspond to the field linewidth as obtained in the conventional FMR measurement. We briefly discuss below the connection between the field linewidth and frequency linewidth. As we are mainly discussing the ferromagnetic thin film and nanostructures, so the Kittel expression (Eq. 4.37) in the presence of uniaxial anisotropy field, which is occasionally encountered while studying them, is given as:

$$f_{Kittel}(H_{ext}) = \frac{|\gamma|}{2\pi}\mu_0\sqrt{(H_{ext}+H_K)(H_{ext}+H_K+M_S)} \qquad (4.37)$$

By replacing $f_{Kittel}(H_{ext})$ by f and H_{ext} by $H_{Kittel}(f)$ and subsequently solving for Eq. 4.37, one may obtain the inverse connection for the Kittel FMR field $H_{Kittel}(f)$ for any specific frequency. It is worth mentioning here, from the Kittel equation it is evident that the FMR frequency nonlinearly scales with FMR field. This nonlinear scaling primarily originates due to the ellipticity of the FM precession cone given by the square root of the ratio of the two factors inside the square root in Eq. 4.37. Interestingly, due to this ellipticity, a nonlinear relation between the field swept linewidth ΔH for a given frequency f and resonance field $H_{Kittel}(f)$ is produced. Also, this ellipticity affects the frequency swept linewidth Δf for a given field H_{ext} and resonance frequency $f_{Kittel}(H_{ext}) = f$. For the case of relatively small linewidth in comparison with the FMR field or frequency, one can obtain field swept and frequency swept linewidth relation by using the Kittel Eq. 4.37 in combination with a simple differentiation as expressed in Eqs. 4.38 and 4.39:

$$\Delta f = \Delta H \frac{\partial f_{Kittel}(H_{ext})}{\partial H_{ext}}\bigg|_{H_{ext}=H_{Kittel}(f)} = |\gamma|P_A(f)\Delta H \qquad (4.38)$$

and

$$\Delta H = \Delta f \frac{\partial H_{ext}(f)}{\partial f}\bigg|_{f=f_{Kittel}(H_{ext})} = \frac{\Delta f}{|\gamma|P_A(f)} \qquad (4.39)$$

where $P_A(f)$ is a dimensionless factor which accounts for the ellipticity of the FMR response in relaxation rate and linewidth analyses. For an in-plane magnetized film, the $P_A(f)$ function is reduced to the following form:

$$P_A(f) = \sqrt{1+\left(\frac{|\gamma|\mu_0 M_S}{4\pi f}\right)^2} \qquad (4.40)$$

Thus, using Eq. 4.40, the frequency linewidth is expressed as:

$$\Delta f = (|y|\Delta H_0 + 4\pi\alpha f)\sqrt{1+\left(\frac{|\gamma|\mu_0 M_S}{4\pi f}\right)^2} \qquad (4.41)$$

We would like to emphasize here that the connection between the FMR field linewidth and frequency linewidth is quite non-trivial. Though one may expect the frequency linewidth to increase with the increase in frequency, similar to the case for the field linewidth, however, occasionally the frequency linewidth is found to decrease with the increase in frequency. Ultimately, as the linewidth in both cases is related to the extremely important parameter, i.e. magnetic damping, hence, a careful interpretation of frequency linewidth is necessary for various systems in order to extract the damping information. We refer the readers to other existing literatures for detailed discussion of this topic as it is beyond the scope of the present book.

It is worth to briefly refer here that VNA-FMR technique has been extremely useful in investigating the collective magnetization dynamics in magnetic nanostructures. Spin-wave modes in periodically patterned magnetic nanostructures popularly known as magnonic crystals have been investigated by a number of groups using VNA-FMR technique. In 2005, Giesen et al. [64] studied an array of $Ni_{80}Fe_{20}$ nanorings (750 nominally identical rings with outer diameter 2.0 μm, thickness 30 nm, and width 250 nm) using VNA-FMR. Using magnetic force microscopy, they inferred the presence of vortex state and onion state for small applied bias magnetic field. Note that in order to access the characteristic spin configuration in nanorings, it was important to use VNA-FMR as a conventional FMR experiment with a resonator cavity with a fixed frequency drives the resonance condition by sweeping the bias magnetic field which may destroy the characteristic spin configuration. The experimental setup consisted of micron-sized CPW connected to a VNA with a frequency range 45 MHz–20 GHz. The nanoring array was directly prepared on the central conductor of the broadband waveguide using e-beam lithography. In their measurement, they observed a series of FMR modes originating from localized spin waves. Interestingly, these modes were found to show hysteretic behavior and irreversible jumps when the rings are switched between the characteristic spin configurations, i.e. between the onion and the vortex state.

Neusser et al. in 2008 [65] performed spin-wave spectroscopy study in $Ni_{80}Fe_{20}$ square antidot lattice with a lattice constant of 490 nm using VNA-FMR by varying the tilt angle of the applied magnetic field with respect to the lattice axis. The circular antidots with a diameter of 240 nm were prepared using focused ion beam. Interestingly, in this study, they found coexistence of four different prominent spin-wave modes for certain range of tilt angle. An important manifestation of this work was the control of standing and propagating spin waves via the field orientation and strength in antidot lattice which could be significantly important in the field of magnonics.

A custom-built broadband VNA-FMR setup developed at S N Bose Centre has been successfully applied for addressing numerous interesting issues related to spin dynamics in arrays of ferromagnetic nanodots with different shape and size along with their arrangement in various lattice symmetries. In 2015, Choudhury et al. [66] reported the spin-wave dynamics in interesting 2-D arrays of $Ni_{80}Fe_{20}$-filled antidots embedded in continuous $Co_{50}Fe_{50}$ film with two different antidot shapes, namely circular and square, arranged in a square lattice (bicomponent magnonic crystal). Periodic array of these structures were fabricated using delicate e-beam lithography process to ensure that at the interface both the constituent materials are exchange coupled. The width of each $Ni_{80}Fe_{20}$-filled region (antidot) was ~400 nm with maximum ±5% deviation, and the edge-to-edge separation between them was 600 nm with maximum ±5% deviation so that both the magnetostatic interaction and the exchange interaction may influence the spin-wave spectra. For the broadband FMR measurement, a CPW of gold (thickness ~150 nm) was deposited on top of the array structures. In their measurement, they observed an asymmetry in the dispersion of spin-wave frequency with bias field in case of filled antidot lattices with both circular and square shapes as opposed to their unfilled counterparts. Most importantly, a minimum in the frequency occurred at a negative bias field which confirmed the presence of interelement exchange interaction at the $Ni_{80}Fe_{20}$ and $Co_{50}Fe_{50}$ interface. A significant variation in the spin-wave spectra was observed, as the embedded element shape varies from circle to square. All the spin-wave modes observed in the experiment were analyzed using micromagnetic simulations, and internal magnetic field was found to play important role. Furthermore, spin-wave frequencies and bandgaps showed bias field tunability which is crucial for application.

Due to limitation in space, only few limited studies have been discussed here using the experimental techniques described above. In order to get detailed information on these, we refer to existing literatures for interested readers [67–69].

4.4 Advantage and Disadvantage of Various Methods

Here, we briefly discuss the advantages and disadvantages associated with investigating magnetization dynamics using various experimental techniques. A great number of important insights into the spin dynamics have been obtained from magnetic measurements using TR-MOKE, BLS, and FMR and their different variants. The type of information one needs to extract from the spin dynamics investigation of a particular system governs the experimental technique to be applied in the study.

The TR-MOKE method offers an extremely high temporal resolution limited only by the laser pulse width and can resolve ultrafast magnetization dynamics, including ultrafast demagnetization, relaxation mechanisms, and various nonlinear effects as well as the carrier and phonon dynamics. It is extremely advantageous technique to measure damping directly in the time domain from the decaying precessional profile. Also, TR-MOKE method is highly localized due to the focused laser spot, which avoids problems of variations and inhomogeneities from larger area averaging. For the TR-MOKE, the probed area can be much smaller ($\sim 1 \ \mu m^2$ or lower) which is significantly smaller than the probe area in frequency domain-based techniques [70]. On this length scale, the magnetic properties of the sample are more homogeneous and thus the effect of linewidth broadening due to inhomogeneities is weaker. Such localized measurements can thus give a better representation of the local damping and hence the local physical environment which affects the spin dynamics. Furthermore, TR-MOKE can also be used to map the variations in damping behavior across a sample. A further advantage is that the modal composition of the magnetization oscillations can be observed in the time domain and the damping for each mode can be assessed. In addition to damping measurements, TR-MOKE technique has been used to study precessional switching in ferromagnetic thin films and patterned structures and the switching time can be precisely determined from the time domain signal. Also, it has been extensively used to study the local precessional dynamics in arrays of nanomagnets and even in single nanomagnets well beyond the optical diffraction limit. All-optical TR-MOKE can measure samples in any form without the requirement of fabrication of complicated device structures for excitation of spin waves. Recently, it has also been used to efficiently detect various spin–orbit effects and interface effects. The disadvantages of this method are that it involves expensive equipment and the optical alignment is very sensitive and non-trivial. This has partly been overcome by the advent of a benchtop TR-MOKE microscope [25], which is cheaper and more stable but lacks the high temporal resolution of a more involved femtosecond laser-based TR-MOKE microscope.

The primary advantage of Brillouin light scattering is its wave-vector sensitivity, which allows the measurement of spin-wave frequency versus wave vector dispersion in different geometry in ferromagnetic thin films, multilayers, and patterned nanostructures. In particular, for patterned one- and two-dimensional ferromagnetic nanostructures, the so-called magnonic crystal [68], it gives us the valuable information about magnon bandgap and group velocity. One of the great advantages of BLS over all competing techniques is its remarkable sensitivity down to monolayer of magnetic material thickness. Most interestingly it allows the detection of thermally activated, incoherent spin waves in systems without any external excitation. It is worth mentioning here, this independence of the BLS technique from antenna structures for the external excitation enables the flexible realization of different measurement geometries to probe spin dynamics in a wide wave vector as well as spectral range. Microfocused BLS has the ability to image the propagating spin wave which provides important insights to the operation of novel spin-wave-based devices such as spin-wave logic devices. The disadvantage of this method is also its high cost and sensitive optical alignment procedure. Measurement of thermal magnon is also very slow.

References

1. Kirilyuk A, Kimel AV, Rasing T (2010) Ultrafast optical manipulation of magnetic order. Rev Mod Phy. 82(3):2731–2784. doi:10.1103/RevModPhys.82.2731
2. Barman A, Haldar A (2014) Time-domain study of magnetization dynamics in magnetic thin films and micro-and nanostructures. In: Camley RE and Stamps RL (ed) Solid State Physics, vol 65, pp 1–108. Elsevier doi:10.1016/B978-0-12-800175-2.00001-7
3. Kittel C (1948) On the theory of ferromagnetic resonance absorption. Phy Rev 73(2):155–161. doi:10.1103/PhysRev.73.155

4. Denysenkov VP, Grishin AM (2003) Broadband ferromagnetic resonance spectrometer. Rev Sci Instrum 74(7):3400–3405. doi:10.1063/1.1581395
5. Brillouin L (1922) Diffusion of light and X-rays by a transparent homogeneous body. Ann Phys 17:88
6. Demokritov SO, Hillebrands B, Slavin AN (2001) Brillouin light scattering studies of confined spin waves: linear and nonlinear confinement. Phy Rep 348(6):441–489. doi:10.1016/S0370-1573(00)00116-2
7. Silva TJ, Lee CS, Crawford TM, Rogers CT (1999) Inductive measurement of ultrafast magnetization dynamics in thin-film permalloy. J Appl Phy 85(11):7849–7862. doi:10.1063/1.370596
8. Freeman MR, Brady MJ, Smyth J (1992) Extremely high frequency pulse magnetic resonance by picosecond magneto-optic sampling. Appl Phy Lett 60(20):2555–2557. doi:10.1063/1.106911
9. van Kampen M, Jozsa C, Kohlhepp JT, LeClair P, Lagae L, de Jonge WJM, Koopmans B (2002) All-optical probe of coherent spin waves. Phys Rev Lett 88(22):227201. doi:10.1103/PhysRevLett.88.227201
10. Hiebert WK, Stankiewicz A, Freeman MR (1997) Direct observation of magnetic relaxation in a small permalloy disk by time-resolved scanning Kerr microscopy. Phys Rev Lett 79 (6):1134–1137. doi:10.1103/PhysRevLett.79.1134
11. Barman A, Wang SQ, Maas JD, Hawkins AR, Kwon S, Liddle A, Bokor J, Schmidt H (2006) Magneto-optical observation of picosecond dynamics of single nanomagnets. Nano Lett 6 (12):2939–2944. doi:10.1021/nl0623457
12. Kerr J (1877) On rotation of the plane of polarization by reflection from the pole of a magnet. Philos Mag Ser 3(19):321–343. doi:10.1080/14786447708639245
13. Hulme HR (1932) The Faraday effect in ferromagnetics. Proc R Soc Lond Ser A 135 (826):237–257. doi:10.1098/rspa.1932.0032
14. Argyres PN (1955) Theory of the Faraday and Kerr effects in ferromagnetics. Phys Rev 97 (2):334–345. doi:10.1103/PhysRev.97.334
15. Roth LM (1964) Theory of the Faraday effect in solids. Phys. Rev 133(2A):A542–A553. doi:10.1103/PhysRev.133.A542
16. Kubo R (1956) A general expression for the conductivity tensor. Can J Phys 34(12A):1274–1277. doi:10.1139/p56-140
17. Koopmans B (2002) Laser induced magnetization dynamics. In: Hillebrands B, Ounadjela K (eds) Spin Dynamics in Confined Magnetic Structures II. Springer, New York, pp 253–312
18. Razdolski I, Alekhin A, Martens U, Bürstel D, Diesing D, Münzenberg M, Bovensiepen U, Melnikov A (2017) Analysis of the time-resolved magneto-optical Kerr effect for ultrafast magnetization dynamics in ferromagnetic thin films. J Phys Condens Matter 29(17):174002. doi:10.1088/1361-648X/aa63c6
19. Zhang GP, Hübner W (2000) Laser-induced ultrafast demagnetization in ferromagnetic metals. Phys Rev Lett 85(14):3025–3028. doi:10.1103/PhysRevLett.85.3025
20. Dietrich W, Proebster WE (1960) Millimicrosecond magnetization reversal in thin magnetic films. J Appl Phys 31(5):S281–S282. doi:10.1063/1.1984700
21. Wolf P (1961) Free oscillations of the magnetization in permalloy films. J Appl Phys 32(3): S95–S96. doi:10.1063/1.2000514
22. Freeman MR, Ruf RR, Gambino RJ (1991) Picosecond pulsed magnetic fields for studies of ultrafast magnetic phenomena. IEEE Trans Magn 27(6):4840–4842. doi:10.1109/20.278964
23. Koopmans B (2003) Laser induced magnetization dynamics. In: Hillebrands B, Ounadjela K (eds) Spin Dynamics in Confined Magnetic Structures II. Springer, New York, pp 253–312
24. Bigot JY, Guidoni L, Beaurepaire E, Saeta PN (2004) Femtosecond spectrotemporal magneto-optics. Phys Rev Lett 93(7):077401. doi:10.1103/PhysRevLett.93.077401
25. Barman A, Kimura T, Otani Y, Fukuma Y, Akahane K, Meguro S (2008) Benchtop time-resolved magneto-optical Kerr magnetometer. Rev Sci Instrum 79(12):123905. doi:10.1063/1.3053353

26. Barman A, Barman S, Kimura T, Fukuma Y, Otani Y (2010) Gyration mode splitting in magnetostatically coupled magnetic vortices in an array. J Phys D Appl Phys 43(42):422001. doi:10.1088/0022-3727/43/42/422001
27. Stotz JAH, Freeman MR (1997) A stroboscopic scanning solid immersion lens microscope. Rev Sci Instrum 68(12):4468–4477. doi:10.1063/1.1148416
28. Park JP, Eames P, Engebretson DM, Berezovsky J, Crowell PA (2002) Spatially resolved dynamics of localized spin-wave modes in ferromagnetic wires. Phys Rev Lett 89 (27):277201. doi:10.1103/PhysRevLett.89.277201
29. Neudert A, Keatley PS, Kruglyak VV, McCord J, Hicken RJ (2008) Excitation and imaging of precessional modes in soft-magnetic squares. IEEE Trans Magn 44(11):3083–3086. doi:10. 1109/tmag.2008.2001653
30. Barman A, Kruglyak VV, Hicken RJ, Rowe JM, Kundrotaite A, Scott J, Rahman M (2004) Imaging the dephasing of spin wave modes in a square thin film magnetic element. Phys Rev B 69(17):174426. doi:10.1103/PhysRevB.69.174426
31. Barman A, Sharma RC (2007) Micromagnetic study of picosecond dephasing of spin waves in a square magnetic element. J Appl Phys 102(5):053912. doi:10.1063/1.2776233
32. Raman CV (1928) A change of wave-length in light scattering. Nature 121:619. doi:10.1038/ 121619b0
33. Gammon PH, Kiefte H, Clouter MJ (1983) Elastic constants of ice samples by Brillouin spectroscopy. J Phys Chem 87(21):4025–4029. doi:10.1021/j100244a004
34. Li F, Cui Q, He Z, Cui T, Zhang J, Zhou Q, Zou G, Sasaki S (2005) High pressure-temperature Brillouin study of liquid water: evidence of the structural transition from low-density water to high-density water. J Chem Phys 123(17):174511. doi:10.1063/1. 2102888
35. Courtens E, Pelous J, Phalippou J, Vacher R, Woignier T (1987) Brillouin-scattering measurements of phonon-fracton crossover in silica aerogels. Phys Rev Lett 58(2):128–131. doi:10.1103/PhysRevLett.58.128
36. Reiß S, Burau G, Stachs O, Guthoff R, Stolz H (2011) Spatially resolved Brillouin spectroscopy to determine the rheological properties of the eye lens. Biomed Opt Express 2 (8):2144–2159. doi:10.1364/boe.2.002144
37. Mandelstam LI (1926) Light scattering by inhomogeneous media. Zh Russ Fiz-Khim Ova 58:381
38. Gross E (1930) Change of wave-length of light due to elastic heat waves at scattering in liquids. Nature 126:201–202. doi:10.1038/126201a0
39. Sandercock JR (1971) Paper presented at the second international conference on light scattering in solids, Flammarion, Paris
40. Mock R, Hillebrands B, Sandercock R (1987) Construction and performance of a Brillouin scattering set-up using a triple-pass tandem Fabry-Perot interferometer. J Phys E: Sci Instrum 20(6):656. doi:10.1088/0022-3735/20/6/017
41. Wettling W, Smith RS, Jantz W, Ganser PM (1982) Single crystal Fe films grown on GaAs substrates. J Magn Magn Mater 28(3):299–304. doi:10.1016/0304-8853(82)90063-4
42. Sandercock J, Wettling W (1978) Light scattering from thermal magnons in Iron and Nickel. IEEE Trans Magn 14(5):442–444. doi:10.1109/tmag.1978.1059895
43. Sandercock J (1999) Operator manual for the tandem Fabry-Perot interferometer
44. Demidov VE, Demokritov SO, Hillebrands B, Laufenberg M, Freitas PP (2004) Radiation of spin waves by a single micrometer-sized magnetic element. Appl Phys Lett 85(14):2866–2868. doi:10.1063/1.1803621
45. Gubbiotti G, Carlotti G, Madami M, Tacchi S, Vavassori P, Socino G (2009) Setup of a new brillouin light scattering apparatus with submicrometric lateral resolution and its application to the study of spin modes in nanomagnets. J Appl Phys. 105(7):07D521. doi:10.1063/1. 3068428
46. Serga AA, Schneider T, Hillebrands B, Demokritov SO, Kostylev MP (2006) Phase-sensitive Brillouin light scattering spectroscopy from spin-wave packets. Appl Phys Lett 89(6):063506. doi:10.1063/1.2335627

47. Vogt K, Schultheiss H, Hermsdoerfer SJ, Pirro P, Serga AA, Hillebrands B (2009) All-optical detection of phase fronts of propagating spin waves in a $Ni_{81}Fe_{19}$ microstripe. Appl Phys Lett 95(18):182508. doi:10.1063/1.3262348

48. Fohr F, Serga AA, Schneider T, Hamrle J, Hillebrands B (2009) Phase sensitive Brillouin scattering measurements with a novel magneto-optic modulator. Rev Sci Instrum 80 (4):043903. doi:10.1063/1.3115210

49. Bauer M, Büttner O, Demokritov SO, Hillebrands B, Grimalsky V, Rapoport Y, Slavin AN (1998) Observation of spatiotemporal self-focusing of spin waves in magnetic films. Phys Rev Lett 81(17):3769–3772. doi:10.1103/PhysRevLett.81.3769

50. Serga AA, Demokritov SO, Hillebrands B, Slavin AN (2004) Self-generation of two-dimensional spin-wave bullets. Phys Rev Lett 92(11):117203. doi:10.1103/PhysRevLett.92.117203

51. Schultheiss H, Sandweg CW, Obry B, Hermsdörfer S, Schäfer S, Leven B, Hillebrands B (2008) Dissipation characteristics of quantized spin waves in nano-scaled magnetic ring structures. J Phys D Appl Phys 41(16):164017. doi:10.1088/0022-3727/41/16/164017

52. Haldar A, Banerjee C, Laha P, Barman A (2014) Brillouin light scattering study of spin waves in NiFe/Co exchange spring bilayer films. J Appl Phys 115(13):133901. doi:10.1063/1.4870053

53. Banerjee C, Chaurasiya AK, Saha S, Sinha J, Barman A (2015) Tunable spin wave properties in $[Co/Ni_{80}Fe_{20}]r$ multilayers with the number of bilayer repetition. J Phys D-Appl Phys 48 (39):395001. doi:10.1088/0022-3727/48/39/395001

54. Banerjee C, Pal S, Ahlberg M, Nguyen TNA, Akerman J, Barman A (2016) All-optical study of tunable ultrafast spin dynamics in Co/Pd/NiFe systems: the role of spin-twist structure on Gilbert damping. RSC Adv 6(83):80168–80173. doi:10.1039/c6ra12227b

55. Banerjee C, Loong LM, Srivastava S, Pal S, Qiu XP, Yang H, Barman A (2016) Improvement of chemical ordering and magnetization dynamics of Co-Fe-Al-Si Heusler alloy thin films by changing adjacent layers. RSC Adv 6(81):77811–77817. doi:10.1039/c6ra05535d

56. Sinha J, Banerjee C, Chaurasiya AK, Hayashi M, Barman A (2015) Improved magnetic damping in CoFeB|MgO with an N-doped Ta underlayer investigated using the Brillouin light scattering technique. RSC Adv 5(71):57815–57819. doi:10.1039/c5ra06925d

57. Chaurasiya AK, Banerjee C, Pan S, Sahoo S, Choudhury S, Sinha J, Barman A (2016) Direct observation of interfacial Dzyaloshinskii-Moriya interaction from asymmetric spin-wave propagation in $W/CoFeB/SiO_2$ heterostructures down to sub-nanometer CoFeB thickness. Sci Rep 6:32592. doi:10.1038/srep32592

58. Griffiths JHE (1946) Anomalous high-frequency resistance of ferromagnetic metals. Nature 158(4019):670

59. Bady I (1967) Measurement of linewidth of single crystal ferrites by monitoring the reflected wave in short-circuited transmission line. IEEE Trans Magn 3(3):521–526. doi:10.1109/tmag.1967.1066105

60. Kalarickal SS, Krivosik P, Wu M, Patton CE, Schneider ML, Kabos P, Silva TJ, Nibarger JP (2006) Ferromagnetic resonance linewidth in metallic thin films: comparison of measurement methods. J Appl Phys 99(9):093909. doi:10.1063/1.2197087

61. Michael F (1998) Ferromagnetic resonance of ultrathin metallic layers. Rep Prog Phys 61 (7):755. doi:10.1088/0034-4885/61/7/001

62. Celinski Z, Heinrich B (1991) Ferromagnetic resonance linewidth of Fe ultrathin films grown on a bcc Cu substrate. J Appl Phys 70(10):5935–5937. doi:10.1063/1.350110

63. Barry W (1986) A broad-band, automated, stripline technique for the simultaneous measurement of complex permittivity and permeability. IEEE Trans Microw Theory Tech 34(1):80–84. doi:10.1109/tmtt.1986.1133283

64. Giesen F, Podbielski J, Korn T, Steiner M, van Staa A, Grundler D (2005) Hysteresis and control of ferromagnetic resonances in rings. Appl Phys Lett 86(11):112510. doi:10.1063/1.1886247

65. Neusser S, Botters B, Becherer M, Schmitt-Landsiedel D, Grundler D (2008) Spin-wave localization between nearest and next-nearest neighboring holes in an antidot lattice. Appl Phys Lett 93(12):122501. doi:10.1063/1.2988290
66. Choudhury S, Saha S, Mandal R, Barman S, Otani Y, Barman A (2016) Shape- and interface-induced control of spin dynamics of two-dimensional bicomponent magnonic crystals. ACS Appl Mater Interfaces 8(28):18339–18346. doi:10.1021/acsami.6b04011
67. Neusser S, Grundler D (2009) Magnonics: spin waves on the nanoscale. Adv Mater 21 (28):2927–2932. doi:10.1002/adma.200900809
68. Krawczyk M, Grundler D (2014) Review and prospects of magnonic crystals and devices with reprogrammable band structure. J Phys: Condens Matter 26(12):123202. doi:10.1088/0953-8984/26/12/123202
69. Mahato BK, Choudhury S, Mandal R, Barman S, Otani Y, Barman A (2015) Tunable configurational anisotropy in collective magnetization dynamics of $Ni_{80}Fe_{20}$ nanodot arrays with varying dot shapes. J Appl Phys 117(21):213909. doi:10.1063/1.4921976
70. Neudecker I, Woltersdorf G, Heinrich B, Okuno T, Gubbiotti G, Back CH (2006) Comparison of frequency, field, and time domain ferromagnetic resonance methods. J Magn Magn Mater 307(1):148–156. doi:10.1016/j.jmmm.2006.03.060

Chapter 5
Factors Affecting Spin Dynamics

As discussed in previous chapters, the precessional magnetization dynamics is described by Landau–Lifshitz–Gilbert (LLG) equation. It may be noticed from the LLG equation that four important material parameters, namely, Landé g-factor, saturation magnetization, magnetic anisotropy, and Gilbert damping parameters directly govern the precessional dynamics. Ultrafast demagnetization and relaxation phenomena are also influenced by material parameters and sample conditions. To understand deeply, various factors related to material aspects of the specimen which affect the dynamics, numerous interesting theoretical and experimental studies have been performed [1–3]. Due to the vastness and abstract nature of this topic, we try to provide only a glimpse of the factors responsible in terms of the choice of the materials in modifying the spin dynamics. As discussed in previous chapters, various interesting measurement techniques evolved with time to measure the spin dynamics. Also, various materials, for example, conventional magnetically ordered materials (ferromagnet [4, 5] and ferrimagnet [2]), magnetic half metals [6], magnetic insulators [7], magnetic dielectrics, etc., have been extensively studied [8]. For the sake of in-depth understanding, we will mainly concentrate on pump-probe based measurement technique, specifically, time-resolved magneto-optical Kerr effect technique, and we will discuss the important studies which have helped to relate the properties of materials that affect the spin dynamics. For the specimen, metallic ferromagnets are considered as important and complex systems to understand because of the itinerant magnetism and huge application potential ranging from power transformers, data storage, and modern spintronics. We will also specifically focus on ferromagnetic thin films, multilayers, and nanostructures where the role of shape, crystallinity, and interface become important.

A. Barman and J. Sinha, *Spin Dynamics and Damping in Ferromagnetic Thin Films and Nanostructures*, https://doi.org/10.1007/978-3-319-66296-1_5

5.1 Material Properties

Material properties play a crucial role in governing the magnetization dynamics and spin dynamics in wide range of materials, for example, Ni, Co, Fe, Gd, alloys such as NiFe, CoFe, CoFeB, GdFeCo, TbCoFe, CoPt, FePt, half-metallic Heusler alloys with high T_C, magnetic garnet films, etc., have been thoroughly studied. Agranat et al., in 1984 [9], and Vaterlaus et al. [10], in 1990, made early attempts to perform ultrafast time-resolved studies using picosecond laser pulses with an aim to investigate the impact of laser pulses on the magnetization of Ni and Fe. However, these studies were not conclusive in observing any magnetic effects due to the fact that the available laser pulses were of the same duration or even longer (few ns) than the relevant time scales of spin dynamics. It was realized later that the various parts of the system remained in equilibrium with each other over such long time scale and hence, the system followed the excitation profile for such long excitations. In 1991, Vaterlaus et al. [11], using time-resolved spin-polarized photoemission as a probe of the magnetization, succeeded in estimating the spin–lattice relaxation time in Gd films and found it to be $\sim 100 \pm 80$ ps. As discussed in earlier chapters, it is worth to recall here that the ultrafast demagnetization was reported in Ni films by Beaurepaire et al. in 1996 [4]. Experimental studies of sub-picosecond mag- netization dynamics were subsequently carried out in Fe by Kampfrath et al. in 2002 [12] and Carpene et al. in 2008 [13]. Koopmans et al. [3, 14, 15] invoked microscopic three-temperature model to explain the drastically different demagne- tization times of Gd and for conventional ferromagnet Fe, Co, Ni as well as the detailed dynamics of the demagnetization and its recovery. These studies indicated the fact that the intrinsic time scales involved with the magnetization dynamics in Gd and conventional ferromagnets differ as they originate from different mechanisms.

Furthermore, in 2002, Guidoni et al. [5] investigated magnetization dynamics in $CoPt_3$ thin film using laser pulses as short as 20 fs duration. By carefully separating the dynamics of the diagonal and the off-diagonal elements of the time dependent dielectric tensor, it was shown that a significant demagnetization can be obtained at a sub-100 fs time scale as reproduced in Fig. 5.1. It may be noticed from Fig. 5.1a, for short temporal delays $t \leq 150$ fs, the time evolution of the real and imaginary parts of complex polarization rotation ($\Theta_F = \theta_F + i\eta_F$) is strongly different. Similar observations were also made by Koopmans et al. in an earlier work in 2000 [14]. In Fig. 5.1b, the real and imaginary parts of the ratio of the off-diagonal and diagonal elements of the first order dielectric tensor ($\tilde{Q} = q\prime + iq\prime\prime$) are plotted which also show different behavior. This study led to an important finding that the ultrafast spin dynamics occurs during the thermalization of the electronic populations with a characteristic time of about 50 fs and this process is followed by a quasistatic equilibrium where the spins follow the dynamics of the electronic temperature [5]. Moreover, in 2003, Rhie et al. [16] showed by time-resolved photoemission study that the exchange splitting between majority and minority spin bands is affected at a similar time scale. Using *ab initio* calculation, Oppeneer and Liebsch in 2004 [17]

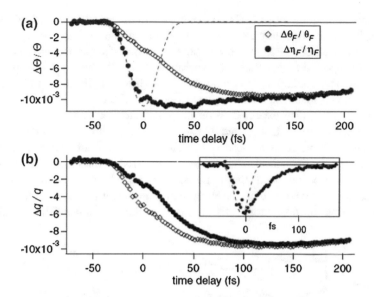

Fig. 5.1 a Time-resolved Faraday MO signals $\Delta\theta/\theta$ ($\Diamond$) and $\Delta\eta/\eta$ ($\bullet$). The pump-probe cross correlation (dashed line) is displayed for reference. **b** Short-delay relative variations of $\Delta q'/q'$ ($\Diamond$) and $\Delta q''/q''$ ($\bullet$) retrieved from the data in (a). For t $\leq$ 150 fs, the real and the imaginary parts of $\Delta\tilde{Q}$ follow clearly different dynamics. The difference between $\Delta q'/q'$ and $\Delta q''/q''$ is plotted in the inset. *Reprinted with permission from Ref.* [5]. *Copyright 2002 by the American Physical Society*

predicted the breakdown between the magneto-optical response and magnetization in ultrafast pump-probe experiments while theoretically investigating ultrafast demagnetization in Ni. They considered dichroic bleaching and state-blocking effects by evaluating the complex conductivity tensor of Ni for non-equilibrium electron distributions. It was shown that the conductivity tensor, and therefore, the complex Kerr angle can be substantially modified so that the Kerr rotation and ellipticity no longer remain proportional to the magnetization of the sample. From this, it was inferred that the Kerr response at ultrashort time scales may not be considered as a measure of demagnetization. By varying the wavelength of the probe beam in the broad spectral range between 500 and 700 nm, Bigot et al. in 2004 experimentally investigated the role of dichroic bleaching and state-blocking effects in ultrafast magneto-optical pump-probe experiments in $CoPt_3$ [18]. The presence of bleaching and state-blocking effects was expected to result in a spectral dependence in the magneto-optical response associated with the population bleaching. However, in their experiment, the spectral response was found to be rather flat which indicated that the magneto-optical signal predominantly reflects the spin dynamics in this ferromagnet. Due to relatively small magneto-optical signals from Ni, similar measurements in Ni could not be performed and hence the conclusions of Bigot et al. [18] were not enough for general description of magnetization dynamics in other metals such as Ni. Moreover, Cheskis et al. in 2005

[19] observed saturation of ultrafast laser-induced demagnetization at high excitation densities in Ni and they explained it in terms of band filling effects.

In order to further develop the understanding of material parameters, Mann et al. in 2012 [6] reported an interesting comparative study of ultrafast demagnetization for three isoelectronic Heusler compounds (Co_2MnSi, Co_2FeAl, and Co_2MnGe), two Co-Fe-based materials (CoFeGe and CoFeB), and the half-Heusler compound CoMnSb, which is also expected to develop half-metallic features, together with standard Ni, $Ni_{81}Fe_{19}$, CoFe, and CrO_2. The demagnetization curves for the above-mentioned materials as observed in their experiment are shown in Fig. 5.2. In this plot, the demagnetization curves are normalized to the respective maximum demagnetization values and these are shifted for clarity. For reference, the demagnetization curve obtained for Ni in their experiment is plotted at the bottom. It is interesting to note from this figure that the position of the maximum

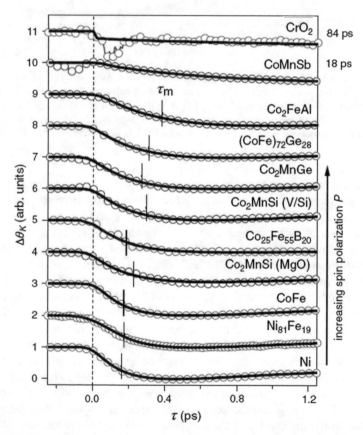

Fig. 5.2 Ultrafast demagnetization spectra. Femtosecond pump-probe experiments for different materials are sorted by their spin polarization P. Heusler compounds are indicated by red data points. The lines are fits using an analytical function derived from a three-temperature model, and the vertical bar marks the value of τ_m. *Reprinted with permission from Ref.* [6]. *Open access material from the American Physical Society*

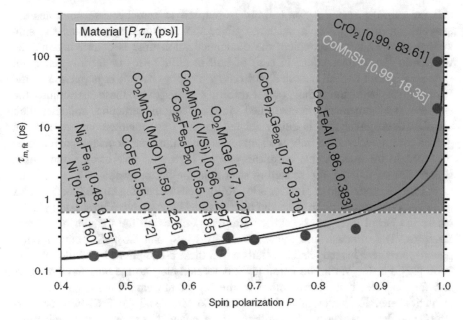

Fig. 5.3 Demagnetization time τ_m versus spin polarization P. The half-metallic properties can be classified in a P versus τ_m plot. If the points lie on top of the lines given, spin-flip blocking is the dominant mechanism describing the simultaneous increase of P and τ_m. The lines are model calculations using Fermi's golden-rule approach showing the $\tau_{el\text{-}sp} \sim (1\text{-}P)^{-1}$ behavior. For each material the demagnetization time τ_m taken from Fig. 5.2 is plotted as a function of the spin polarization P. Note the logarithmic scale for τ_m. Intersections are $P = 0$, $\tau_m = \tau_{el,0}/c^2 = 100$ fs and $P = 1$, $\tau_m = 3$ ps or 1 ns, respectively. Heusler compounds are given in red. In brackets the value of P and τ_m (fs) is given. *Reprinted with permission from authors of Ref. [6]. Open access material from the American Physical Society*

demagnetization of Co_2FeAl is significantly larger in comparison to that of Ni. The extracted value of demagnetization time is plotted against spin polarization for all these materials and it is reproduced in Fig. 5.3. In this figure, the value of spin polarization is mentioned for each material. The spin polarization values for various materials were obtained using spin transport measurements. The choice of these materials allowed these authors to understand the behavior of demagnetization time with spin polarization varying from 45% to almost 100%.

From this study it was understood that at low spin polarization, the demagnetization time is dominated by the fast electron-spin relaxation rate ($\tau_m \sim \tau_{el\text{-}sp}$). The limiting values at $P = 1$ is determined by the lattice–spin relaxation rate $\tau_{lat\text{-}sp}$. In Fig. 5.3, the black and red lines are plotted following model calculations involving Fermi's golden rule approach which indicates $\tau_{el\text{-}sp} \sim (1-P)^{-1}$. The intersection value of black and red curve at $P = 1$ is for $\tau_m = 3$ ps and 1 ns. Particularly, for a half metal, the blocking of spin-flip scattering processes becomes efficient once all the electrons are relaxed below the energy of the half-metallic gap. In such cases, the electron system and the spin system are completely isolated, and the spin-flip scattering probability is reduced to zero because no states are available in one of the spin channels. Since the electronic system and the spin system are thermally decoupled, the demagnetization time is determined by the weak spin–lattice interaction. In case of half-metallic ferro- or ferrimagnets, full spin polarization can be obtained and only one spin channel is populated at the Fermi level with the other one exhibiting a gap. In these materials, the spin-scattering processes are controlled via the electronic structure, and thus their ultrafast demagnetization is solely related to the spin polarization via a Fermi golden rule model. Concurrent with this mechanism, a long demagnetization time observed in the experiments correlates with a high spin polarization due to the suppression of the spin-flip scattering at around the Fermi level.

As the magnetization of Heusler alloy follows Slater Pauling curve [20, 21], thus ultrafast demagnetization in them can be explained by applying a rigid-band picture where the Fermi level can be shifted simply by changing the number of valence electrons. The ultrafast spin dynamics of Co_2MnSi and Co_2FeSi Heusler compounds have been investigated in details under these assumptions by Muller et al. in 2009 [22]. For Co_2MnSi, the Fermi level is located close to the bottom of the gap, and for Co_2FeSi, it is located close to the top of the gap. Thus even for small excitation energies, states of both bands in Co_2MnSi and Co_2FeSi become thermally populated. These states can provide a channel for fast demagnetization. Particularly using numerical simulation, for the Co_2MnSi and Co_2FeSi, the electron and spin dynamics at ultrafast time scale were found to be akin to semiconductors where the excitation and relaxation can be traced to a selected band. For Co_2MnSi, a channel of hot holes below the gap dominates relaxation, whereas for Co_2FeSi, a channel of hot holes above the gap is responsible for relaxation. Interestingly, also for pseudogap material such as partially ordered CoFeGe and CoFeB, it is possible to block spin-flip processes and thus to suppress spin scattering on ultrafast time scales. The relatively large spin polarization in these may be obtained by locally induced partial suppression of states in one spin channel [22].

It is important to mention here that ultrafast laser-induced demagnetization of metals can be accompanied by a change in magnetization reversal behavior. We present here an interesting study by Bigot et al. [23] investigated in 2005, where the influence of anisotropy effects on the magnetization dynamics of ferromagnetic cobalt films and nanoparticles excited with femtosecond laser pulses was carefully investigated. In this study, two different cobalt films, one of thickness 16 nm grown by molecular beam epitaxy on a (0001) oriented sapphire substrate and other of 50 nm grown on (110) MgO substrate, were used. The film grown on sapphire substrate has hexagonal compact crystalline phase with the $\sim$c axis along the (0001) direction (perpendicular to the film) and in-plane Kerr loop saturates for an applied field of 0.9 kOe and it is the easy axis for magnetization. For the film grown on MgO substrate, the resulting $\sim$c axis is in the plane of the sample. The magnetization easy axis is along the in-plane direction and the in-plane Kerr loop saturates at an applied field of 0.2 kOe. In Fig. 5.4, we present the magnetization dynamics results of Bigot et al. as observed in their time-resolved magneto-optical Kerr effect experiments. In Fig. 5.4a, the differential polar signals obtained on the

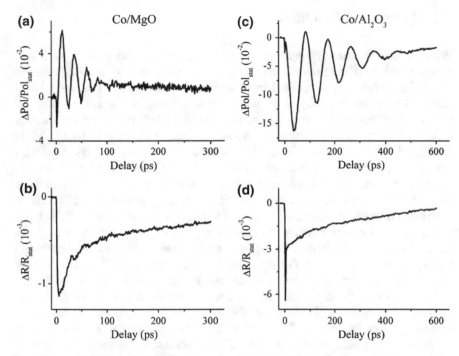

Fig. 5.4 **a** and **b** Differential polar signals and reflectivity for the Co/MgO sample. Corresponding signals for the Co/Al$_2$O$_3$ sample in (**c**) and (**d**). *Reprinted with permission from Ref.* [23] *Copyright 2005 by the Elsevier*

Co/MgO sample for the long temporal delays are shown. The short oscillation period of 24 ps is due to the anisotropy field. The differential reflectivity curve is plotted in Fig. 5.4b, and a comparison with Fig. 5.4a indicates that the precession is damped well before the lattice temperature gets relaxed. Furthermore, it may be noticed that there is additional damping mechanism for the precession which is faster than the thermal diffusion time. Interestingly, for the Co/Al$_2$O$_3$ sample, the precession lasts for much longer duration as seen in the dynamics of the polar signal displayed in Fig. 5.4c compared to the differential reflectivity shown in Fig. 5.4d. Note that the demagnetizing field which opposes to the anisotropy field is perpendicular to the sample in the case of Co/Al$_2$O$_3$. Due to this, the precession period gets significantly enhanced to ∼91 ps. The full recovery of magnetization by the time when the precession gets damped indicates that a single mechanism, i.e., heat diffusion is responsible for the damping of the precession and the complete recovery of the magnetization modulus. The reason for faster damping of precession for the Co/MgO in this study was attributed to larger spin scattering at the metal/substrate interface or due to spin-waves interaction.

5.2   Interfacial Condition

With the development of advanced thin-film deposition system, there have been
tremendous efforts to achieve improved magnetic properties by utilizing interface
effects for the device application. Dramatic effects originating from interface can be
observed by considering layers of material that are only a few atoms thick, and
comprised of multiple layers of different materials. In such cases, the interface can
make up a significant proportion of the total volume of the film. Interestingly, the
interface describes the transition between adjacent layers in the multilayer and has a
structure which can influence the properties within the layers themselves. Following
band theory, overlapping bands with different chemical potentials must join at the
interface, which modifies the electronic and magnetic properties of the material in
this region through hybridization. The interface also contributes energetically in the
system with an energy contribution associated with the interface anisotropy constant. The interface anisotropy may give rise to perpendicular magnetic anisotropy
under appropriate condition. For significantly small ferromagnetic layer thickness,
the surface contribution can be so strong that the energy cost due to the demagnetizing field associated with out-of-plane magnetization can be overcome.
Substantial modification in magnetization dynamics may occur due to the effects
originating from interface. In 2001, Berger [24] showed that the presence of sharp
interfaces in magnetic multilayers causes a local increase of the interaction between
spin waves and conduction electrons and subsequently results in enhanced value of
Gilbert damping parameter near an interface. This also leads to an increased value
of ferromagnetic resonance linewidth in comparison to its value in single layer
films. As the interfaces can dramatically alter the spin-wave lifetimes of all
wavelengths, thus following LLG description, it implies that the interfaces can play
a central role in ultrafast demagnetization by impacting damping at atomistic scales.
Battiato et al. [25] in 2010 claimed an interesting effect related to interface and finite
size effects in ferromagnetic thin-film heterostructures. They found that the mobility
of optically excited carriers can drive spin current at the interfaces which can
provide a non-local mechanism for fast changes of magnetization. In an interesting
recent study, Kuiper et al. (2014) [26] compared the ultrafast demagnetization of
pure Co films with the Co/Pt multilayers. The main aim of their work was to speed
up the demagnetization process by using strong spin–orbit coupling of Co at the
interface of Co/Pt multilayer film. The Co/Pt multilayer in this study was fabricated
using sputter deposition technique on Si substrate coated with SiO_2 and it consists
of $Pt_5/[Co_{0.5}Pt_{0.6}]_{11}/Pt_{1.4}$ where digits represent the thickness in nm. A 15-nm Co
film deposited on MgO substrate was used as reference Co film in their experiment.
The ultrafast demagnetization was investigated for various pump fluence for both
Co and Co/Pt multilayer films, and the maximum demagnetization ($\Delta M_{max}/M_0$) was
found to increase for larger pump fluence. The demagnetization time (τ_m^*) was
estimated using microscopic three-temperature model where transfer of angular
momentum between electron, lattice, and spin is explicitly modeled using
Elliot-Yafet type of spin scattering [27]. Interestingly in their experiment, τ_m^* for

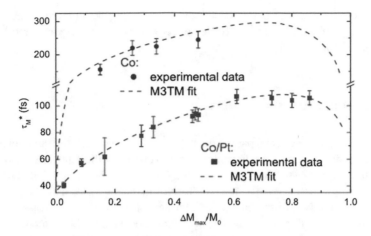

Fig. 5.5 The demagnetization times determined from the individual traces of demagnetization curve in Ref. [26] using the approach of Ref. [27]. The dashed lines represent the results from simulations based on the M3TM. *Reprinted with permission from Ref. [26]. Copyright 2014 by American Institute of Physics*

Co/Pt was found to be 2–3 times less in comparison to that of Co. The variation of τ_m^* as a function of $\Delta M_{max}/M_0$ as observed in this study is reproduced in Fig. 5.5. It clearly indicates that the τ_m^* increases for larger value of $\Delta M_{max}/M_0$, and Co/Pt multilayer has smaller demagnetization time.

Another manifestation of effect of interface on controlling magnetization dynamics can be observed in the study of exchange-coupled NiFe/NiO bilayers by Ju et al. in 1999 [28]. They observed modulation of exchange coupling on a picosecond timescale in ferromagnet/antiferromagnet by comparing the time-resolved magneto-optical response of the NiFe/NiO bilayer with the NiFe thin films without NiO. Interestingly, unpinning of the exchange-bias effect which leads to coherent magnetization rotation in the NiFe film on a time scale of 100 ps was the first reported by these authors. Weber et al. in 2004 [29] investigated the spin dynamics in exchange-coupled NiFe/FeMn using 9 ps laser pulse and found nearly 50% reduction of the exchange-bias field in comparison to its initial value within 20 ps. Moreover, the fast quenching was followed by a slower recovery of the exchange-bias field over a relaxation time of ∼170 ps. Subsequently in 2008 and 2010, Longa et al. [30, 31] investigated the time scale of laser-induced exchange-bias quenching in a polycrystalline Co/IrMn bilayer film and estimated it to be ∼0.7 ps using time-resolved magneto-optical Kerr effect. In this study, the fast decrease in exchange coupling upon laser heating was attributed to spin disorder at the interface.

From the perspective of spin dynamics, the way in which the magnetism changes at an interface is of significant importance, particularly when a ferromagnet forms an interface with heavy metals such as Pt or Pd. Interestingly, in some cases, the ferromagnetism may extend beyond the physical structure of the interface and may result in proximity induced magnetization in the heavy metal layer. The physical

structure of the interface is crucial for the spin-mixing conductivity and the efficiency of the spin–orbit torques. These effects will be specifically covered in more
details in Chap. 6.

5.3 Shape, Size, and Pattern

The effects of shape, size, and pattern on magnetization dynamics are quite interesting
for reduced dimension of the ferromagnetic specimen as the effective fields which
control the magnetic switching process and dynamics can be substantially controlled
via these parameters. Primarily, the equilibrium magnetic states and reversal mechanisms are strongly determined by the interplay of magnetic anisotropies with the
dipole fields which depend on the physical size and shape of the element. From the
application perspective and fundamental understanding of effects of shape, size, and
pattern on spin dynamics, various structures, for example, nanodots, antidots, nanowires, nanorings, etc., and their arrays have been extensively studied by researchers.
Experimental study of magnetization dynamics in magnetic nanostructures and its
arrays was primarily done using FMR and BLS techniques. In 1998, Mathieu et al.
[32] using BLS spectroscopy observed the quantized surface spin waves in periodic
array of micron-sized NiFe wires. Jung et al. in 2002 [33] reported the FMR study of
submicron-sized NiFe dot array. Interestingly, they found that the FMR spectra of the
arrays show a number of additional resonance peaks, whose positions depend strongly
on the orientation of the external magnetic field and the interparticle interaction. These
results were also supported by simulation studies. Overall, the magnetization
dynamics become non-trivial for finite-sized non-ellipsoidal magnetic elements due to
the non-uniformity of internal magnetic field. Spatial confinement and quantization of
spin-wave modes on nanometer length scale may result due to non-uniform demagnetizing field. From the point of view of magnetization dynamics, in case of
non-ellipsoidal magnetic nanodot, the inhomogeneity of the intradot static demagnetization field was also found to be important. BLS measurements in case of arrays of
magnetic nanodots primarily showed standing spin-wave modes of Damon–Eshbach
origin, dipole-exchange mode, and reminiscent of backward-volume mode, in addition to a laterally confined edge mode, with its frequency independent of dot radius.
Following these studies, in the last decade, numerous interesting and fascinating
effects of shape, size, and pattern on spin dynamics have been reported.

Kruglyak et al. [34, 35] in 2005 reported the time-resolved magneto-optical
studies of magnetization dynamics using pulsed field-induced precessional
dynamics in arrays of magnetic nanodots with varying diameter from 630 nm down
to 64 nm. In this study, the $Co_{80}Fe_{20}$ (10 Å)/$Ni_{88}Fe_{12}$ (27 Å) nanodots were fabricated using electron-beam lithography and ion milling. In these arrays, there was
no systematic variation in the interdot spacing and only the effect of size on
magnetization dynamics was studied. Interestingly, the experimentally observed
modes were found to fall upon two branches characterized by two different frequencies, with a crossover from the high- to low-frequency regime as the element

size was reduced to less than 220 nm. Through micro-magnetic simulations, it was confirmed that the magnetization dynamics in these magnetic elements are non-uniform. Furthermore, it was concluded from this study that the interplay between the exchange and demagnetizing fields determines the frequencies and relative amplitudes of the modes in these two branches. For sufficiently small size of the element, a mode confined by the demagnetizing field within the edge regions of the element becomes dominant and the spin dynamics become less uniform. Keatley et al. [36] in 2008 reported a more detailed experimental and simulation study of magnetization dynamics for higher thickness of $Co_{80}Fe_{20}$ (40 Å)/$Ni_{88}Fe_{12}$ (108 Å) with various element size (separation) ranging from 637 nm (25 nm) to 70 nm (37 nm). Larger thickness of the film stack results in non-uniform effective field within the element and static magnetization. Interestingly in this study, above a particular bias field, two branches of excited modes were found to coexist, and below the crossover field, the higher frequency branch disappears. Using micro-magnetic simulations, it was revealed that the higher frequency branch has large mode amplitude at the center of the element in regions of positive effective field, whereas the lower frequency branch has high mode amplitude near the edges of the element perpendicular to the bias field. The complicated evolution of the total effective field within the element mediates the crossover between the higher and lower frequency branches, and moreover, below the crossover region, edge type mode extends into the entire element. Furthermore, it was found using simulation that the majority of the modes are delocalized with finite fast Fourier transform (FFT) magnitude throughout the element. Most interestingly, for nanoscale non-ellipsoidal elements, the delocalized nature of the excited modes appeared to be an intrinsic property. Also in this study, collective nature of the modes with various spatial distributions over the arrays was noticed.

In 2006, Barman et al. studied the dynamics of magnetization of individual Ni disks with 150-nm thickness and diameter varying from 5 μm to 125 nm [37]. They observed a precession of the magnetization with a period that decreases with the diameter of the nanodisks with the effect appearing for sizes below 1 μm and becoming prominent below ~ 300 nm. Due to the appearance of a surface aniso-tropy that competes with the demagnetizing field, for disks of size below 1 μm, the frequency value decreases with the increase in external field. For sizes below 250 nm, the magnetization of the nanostructures undergoes a transition from an in-plane to an out-of-plane direction, and at 125 nm, it converts completely to single-domain. The damping of the same samples also showed a strong size dependence [38]. A pronounced difference in the damping between the micro- and nanomagnets was observed. Microscale magnets showed large damping at low bias fields, whereas nanomagnets exhibited a bias field-independent damping. The observation was interpreted by the interaction of in-plane and out-of-plane preces-sional modes in microscale magnets that result in additional dissipative channels.

In 2011, Rana et al. reported the experimental observation of collective dynamics of square arrays of $Ni_{80}Fe_{20}$ dots with a width of 200 nm as a function of the areal density of the arrays [39]. It was shown that for sufficiently high areal density, the magnetization dynamics displays a single frequency precession which corresponds to a strongly collective behavior of all elements in the array. With decreasing areal density, non-uniform collective modes of the array start to appear. At significantly low areal density, the nanodots behave in decoupled fashion and center and edge modes of the isolated elements dominate. Subsequently, Rana et al. in 2011 reported their seminal work in which 50-nm $Ni_{80}Fe_{20}$ dots with various interparticle distances varying from 50 to 200 nm down to the single nanodot regime was studied [40].

In case of single nanodots, the precessional dynamics showed one dominant resonant edge mode with slightly higher damping than that of the unpatterned thin film. The precession frequency and damping increase monotonically with the increase in the areal density of the array. In comparison to an isolated single nanodot, the precessional dynamics in a coupled nanodot array is found to have different characteristics. A new set of collective modes appears for the case of array, and the interpretation of spin dynamics in terms frequencies of individual nanodots no longer remains valid. In ref. [41], an interesting size-dependent dynamics have been reported in $Ni_{80}Fe_{20}$ square nanodots with 20-nm thickness and with edge width (W) varying between 200 nm and 50 nm. Three distinct size regions were identified. For 200 nm $\leq W \leq$ 150 nm, three clear modes (edge mode (EM), center mode (CM), and a mixed backward volume—Damon Eshbach (BV-DE) mode) are observed while for 100 nm $\leq W \leq$ 75 nm, the BV-DE mode disappeared and only EM and CM appeared. For $W = 50$ nm, another drastic variation is observed with the disappearance of the CM and only the EM becomes prominent as discussed above (Fig. 5.6).

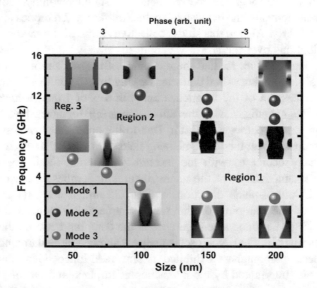

Fig. 5.6 The effect of nanodot size on the precessional modes. Frequencies of the simulated dynamic resonant modes of single $Ni_{80}Fe_{20}$ square dots with varying sizes ($W = 50, 75, 100, 150$ and 200 nm) are plotted as a function of the dot size. The mode (phase) profiles for all modes are shown. The three different background colors show three different regions of interest. Different branches of frequencies are assigned with different colored symbols. The color scale is shown at *top* of the figure

The lattice arrangement plays an important role in the collective precessional dynamics of nanodot arrays. It affects the interdot interaction field and gives rise to the so called extrinsic configurational anisotropy, which is different from the intrinsic configurational anisotropy which arises due to the internal field distribution in confined ferromagnetic elements. A number of works have been reported in the literature on these effects, and recently, Saha et al. [42] have done an extensive work in this field by studying the collective precessional dynamics in circular nanodot arrays arranged in different Bravais (square, rectangular, hexagonal, honeycomb) and non-Bravais (octagonal) lattices. A large variation of the collective spin-wave spectra is observed with the variation in the lattice symmetry (Fig. 5.7), which has been thoroughly interpreted by numerically simulated collective mode profiles and magnetostatic field distribution.

It is important to mention here that the interdot magnetostatic interaction in ordered arrays of nanomagnets can also get affected by the shapes of the elements. Magnetization dynamics can be affected significantly by the shapes of the elements as the profile of the stray magnetic field depends on the shapes of the boundaries of

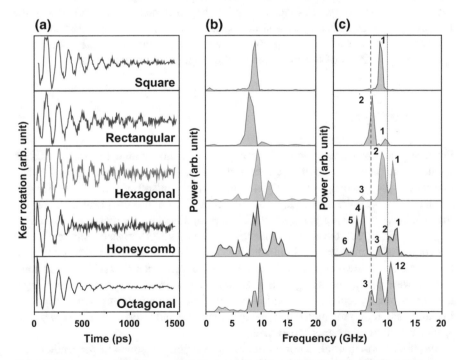

Fig. 5.7 Experimental **a** time-resolved Kerr rotation data and **b** the corresponding FFT power spectra are shown for $Ni_{80}Fe_{20}$ dot lattices with five different lattice symmetry at $H = 1.3$ kOe at in-plane bias magnetic field orientation $\varphi = 0°$. **c** FFT power spectra of the simulated time-domain magnetization for five different lattice symmetry. The mode numbers are shown in the simulated FFT spectra, while the dashed vertical lines show the positions of the center and edge modes of the simulated single dot with width = 100 nm and thickness = 20 nm. *Reproduced with permission from Ref.* [42]. *Copyright 2013 WILEY-VCH Verlag*

the elements as well as the internal magnetic field. In general, the magnetization dynamics of nanomagnets have primarily been studied in square, circular, and ellipsoidal shapes. In square-shaped elements, configurational anisotropies originating from the shapes of the elements have been observed.

Recently, Mahato et al. investigated time-resolved magnetization dynamics in cross-shaped $Ni_{80}Fe_{20}$ nanoelements [43]. The cross shape was of dimension 600 nm in length and width, 20 nm in thickness, and about 50 nm in separation between the nearest edges. Interestingly, in this study, a significant anisotropy in frequencies and nature of the spin-wave modes with applied in-plane bias field angle has been found. Further, Mahato et al. presented a comparative study of magnetization dynamics [44] and configurational anisotropy [45] in a number of non-ellipsoidal elements. The dynamics showed a stark variation with the element shape as the number of modes changed from two to eight from circular- to triangular-shaped elements. The precessional mode profiles changed from conventional EM and CM to a number of standing wave modes of BV and mixed BV-DE origin. Another interesting observation is that while only antisymmetric modes with odd mode numbers appeared for all element shapes, both symmetric and antisymmetric modes were observed in the triangular-shaped element presumably due to the lack of the symmetry in the internal magnetic potential in this element shape.

Overall, the above-described studies provide a broad picture of the influence of spatial characteristics of the materials on the magnetization dynamics.

5.4 Multilayer

From application point of view, it is important to understand the magnetization dynamics in magnetic multilayer. In multilayer, due to the presence of many interfaces, the exchange interaction and surface anisotropy may be strongly enhanced. In case the spacer layer is thin enough to allow coupling between the magnetic layers, the spin-wave spectra change significantly. In magnetic multilayer with ultrathin layers, the dipolar collective modes (except the stack surface mode) get converted into exchange dominated modes (PSSW modes) in the full coupling limit. Primarily, few decades ago, magnetization dynamics investigations in multilayers were performed using BLS technique. The spin-wave spectra in multilayer may additionally contain information on interlayer coupling, magnetic properties including saturation magnetization, g-factor, volume and interface anisotropies, and spatial variation of magnetic parameter.

In 2007, Barman et al. reported the magnetization dynamics in perpendicularly magnetized Co/Pt multilayers by TR-MOKE technique [46]. In this work, a systematic study was performed with $[Co(4\text{Å})/Pt(8\text{Å})]_n$ multilayers with varying number of bilayer repeats (n). Due to large perpendicular magnetic anisotropy (PMA), fast precession frequency was observed for all samples. Interestingly, they found that while the PMA values decrease with increasing n, the damping parameter α increases with increasing n. The enhanced interaction between magnons and conduction electrons was suggested as a possible mechanism for the enhancement of α with n. TR-MOKE measurements on Pt/Co/Pt films showed that the PMA is inversely proportional to the Co layer thickness and α too varied with Co layer thickness without following any specific trend. Sajitha et al. in 2010 reported another interesting study in which CoFeB buffered $[Co(0.3 \text{ nm})/Pd]_6$ multilayer with varying Pd layer thickness was investigated [47]. Specifically, it was found that the α value depends on the Pd layer thickness in this multilayer. The CoFeB buffer layer significantly affects both the precession frequency and α of the films due to the in-plane anisotropy. Later Pal et al. [48], showed a linear relationship between perpendicular magnetic anisotropy and damping α in a series of $[Co(t)/Pd(0.9 \text{ nm})]_8$ multilayer samples with Co layer thickness varying between $0.22 \text{ nm} \leq t \leq 1.0 \text{ nm}$ and explained it in terms of the interfacial d-d hybridization between Co and Pd layers as the primary reason for this observed linear relationship. The debate on the relationship between perpendicular magnetic anisotropy and α still continues.

In multilayers, apart from spin-wave spectra and Gilbert damping, speed and efficiency of ultrafast demagnetization can be also controlled. Malinowski et al. in 2008 [49] pointed out that laser-excited spin-polarized hot electrons could increase and speed up the loss of magnetization in magnetic multilayers. It was shown that efficiency of spin-angular-momentum transfer can be controlled by introducing a spacer layer between two ferromagnetic layers. In particular, if the magnetization in ferromagnetic layers is in parallel configuration, then there will be no spin current, whereas if the magnetization are in antiparallel configuration, then there will be finite spin current. In Co/Pt multilayers, by introducing different spacer layers (NiO or Ru), significant changes in the ultrafast demagnetization were observed. In case of Ru spacer layer, in antiparallel configuration, larger magnetization loss along with $\sim 25\%$ faster demagnetization was observed as compared to the parallel configuration. This difference in the demagnetization times between the parallel and antiparallel configurations was not observed when NiO spacer layer was used; however, there was still a small difference in the amount of demagnetization for parallel and antiparallel configurations. The $\sim 25\%$ decrease in demagnetization time in the antiparallel configuration for Ru spacer layer was attributed to the direct spin-momentum transfer between two Co-Pt layers.

We would like to mention here that there are several important and pioneering studies related to the effect of material aspect on magnetization dynamics, and this topic is vast and not limited to the discussion presented in this chapter.

References

1. Kirilyuk A, Kimel AV, Rasing T (2010) Ultrafast Optical Manipulation of Magnetic Order. Rev Mod Phys 82(3):2731–2784. doi:10.1103/RevModPhys.82.2731
2. Kirilyuk A, Kimel AV, Rasing T (2013) Laser-Induced Magnetization Dynamics and Reversal in Ferrimagnetic Alloys. Rep Prog Phys 76(2):1–35. doi:10.1088/0034-4885/76/2/026501

3. Koopmans B, Ruigrok JJM, Longa FD, de Jonge WJM (2005) Unifying Ultrafast Magnetization Dynamics. Phys Rev Lett 95(26):267207. doi:10.1103/PhysRevLett.95. 267207
4. Beaurepaire E, Merle JC, Daunois A, Bigot JY (1996) Ultrafast Spin Dynamics in Ferromagnetic Nickel. Phys Rev Lett 76(22):4250–4253. doi:10.1103/PhysRevLett.76.4250
5. Guidoni L, Beaurepaire E, Bigot J-Y (2002) Magneto-Optics in the Ultrafast Regime: Thermalization of Spin Populations in Ferromagnetic Films. Phys Rev Lett 89(1):017401. doi:10.1103/PhysRevLett.89.017401
6. Mann A, Walowski J, Münzenberg M, Maat S, Carey MJ, Childress JR, Mewes C, Ebke D, Drewello V, Reiss G, Thomas A (2012) Insights into Ultrafast Demagnetization in Pseudogap Half-Metals. Phys Rev X 2(4):041008. doi:10.1103/PhysRevX.2.041008
7. Hansteen F, Kimel A, Kirilyuk A, Rasing T (2006) Nonthermal Ultrafast Optical Control of the Magnetization in Garnet Films. Phys Rev B 73(1) doi: 01442110.1103/PhysRevB.73.014421
8. Vahaplar K, Kalashnikova AM, Kimel AV, Gerlach S, Hinzke D, Nowak U, Chantrell R, Tsukamoto A, Itoh A, Kirilyuk A, Rasing T (2012) All-Optical Magnetization Reversal by Circularly Polarized Laser Pulses: Experiment and Multiscale Modeling. Phys Rev B 85(10) doi:10.1103/PhysRevB.85.104402
9. Agranat MB, Ashitkov SI, Granovskii AB, Rukman GI (1984) Interaction of Picosecond Laser Pulses with the Electron, Spin, and Phonon Subsystems of Nickel. Soviet Physics JETP 59(4):804–806
10. Vaterlaus A, Guarisco D, Lutz M, Aeschlimann M, Stampanoni M, Meier F (1990) Different Spin and Lattice Temperatures Observed by Spin-Polarized Photoemission with Picosecond Laser Pulses. J Appl Phys 67(9):5661–5663. doi:10.1063/1.345918
11. Vaterlaus A, Beutler T, Meier F (1991) Spin-Lattice Relaxation Time of Ferromagnetic Gadolinium Determined with Time-Resolved Spin-Polarized Photoemission. Phys Rev Lett 67(23):3314–3317. doi:10.1103/PhysRevLett.67.3314
12. Kampfrath T, Ulbrich RG, Leuenberger F, Münzenberg M, Sass B, Felsch W (2002) Ultrafast Magneto-Optical Response of Iron Thin Films. Phys Rev B 65(10):104429. doi:10.1103/PhysRevB.65.104429
13. Carpene E, Mancini E, Dallera C, Brenna M, Puppin E, De Silvestri S (2008) Dynamics of Electron-Magnon Interaction and Ultrafast Demagnetization in Thin Iron Films. Phys Rev B 78(17):174422. doi:10.1103/PhysRevB.78.174422
14. Koopmans B, van Kampen M, Kohlhepp JT, de Jonge WJM (2000) Ultrafast Magneto-Optics in Nickel: Magnetism or Optics? Phys Rev Lett 85(4):844–847. doi:10.1103/PhysRevLett.85. 844
15. Koopmans B, Malinowski G, Dalla Longa F, Steiauf D, Faehnle M, Roth T, Cinchetti M, Aeschlimann M (2010) Explaining The Paradoxical Diversity of Ultrafast Laser-Induced Demagnetization. Nat Mater 9(3):259–265. doi:10.1038/nmat2593
16. Rhie HS, Dürr HA, Eberhardt W (2003) Femtosecond Electron and Spin Dynamics in Ni/W (110) Films. Phys Rev Lett 90(24):247201. doi:10.1103/PhysRevLett.90.247201
17. Oppeneer PM, Liebsch A (2004) Ultrafast Demagnetization in Ni: Theory Of Magneto-Optics for Non-Equilibrium Electron Distributions. J Phys: Condens Matter 16(30):5519. doi:10. 1088/0953-8984/16/30/013
18. Bigot JY, Guidoni L, Beaurepaire E, Saeta PN (2004) Femtosecond Spectrotemporal Magneto-Optics. Phys Rev Lett 93(7):077401. doi:10.1103/PhysRevLett.93.077401
19. Cheskis D, Porat A, Szapiro L, Potashnik O, Bar-Ad S (2005) Saturation of the Ultrafast Laser-Induced Demagnetization in Nickel. Phys Rev B 72(1):014437. doi:10.1103/PhysRevB.72.014437
20. Slater JC (1937) Electronic Structure of Alloys. J Appl Phys 8(6):385–390. doi:10.1063/1. 1710311
21. Hem CK, Gerhard HF, Claudia F (2007) Calculated Electronic and Magnetic Properties of the Half-Metallic, Transition Metal Based Heusler Compounds. J Phys D Appl Phys 40(6):1507. doi:10.1088/0022-3727/40/6/S01

22. Muller GM, Walowski J, Djordjevic M, Miao G-X, Gupta A, Ramos AV, Gehrke K, Moshnyaga V, Samwer K, Schmalhorst J, Thomas A, Hutten A, Reiss G, Moodera JS, Munzenberg M (2009) Spin Polarization in Half-Metals Probed by Femtosecond Spin Excitation. Nat Mater 8(1):56–61. doi:10.1038/nmat2341
23. Bigot JY, Vomir M, Andrade LHF, Beaurepaire E (2005) Ultrafast Magnetization Dynamics in Ferromagnetic Cobalt: The Role of The Anisotropy. Chem Phys 318(1–2):137–146. doi:10.1016/j.chemphys.2005.06.016
24. Berger L (2001) Effect of Interfaces on Gilbert Damping and Ferromagnetic Resonance Linewidth in Magnetic Multilayers. J Appl Phys 90(9):4632–4638. doi:10.1063/1.1405824
25. Battiato M, Carva K, Oppeneer PM (2010) Superdiffusive Spin Transport as a Mechanism of Ultrafast Demagnetization. Phys Rev Lett 105(2):027203. doi:10.1103/PhysRevLett.105.027203
26. Kuiper KC, Roth T, Schellekens AJ, Schmitt O, Koopmans B, Cinchetti M, Aeschlimann M (2014) Spin-Orbit Enhanced Demagnetization Rate in Co/Pt-Multilayers. Appl Phys Lett 105 (20):202402. doi:10.1063/1.4902069
27. Dalla Longa F., Kohlhepp JT, de Jonge WJM, Koopmans B, Influence of Photon Angular Momentum on Ultrafast Demagnetization in Nickel. Phys Rev B 75(22) (2007). doi:10.1103/ PhysRevB.75.224431
28. Ju G, Nurmikko AV, Farrow RFC, Marks RF, Carey MJ, Gurney BA (1999) Ultrafast Time Resolved Photoinduced Magnetization Rotation in a Ferromagnetic/Antiferromagnetic Exchange Coupled System. Phys Rev Lett 82(18):3705–3708. doi:10.1103/PhysRevLett.82.3705
29. Weber MC, Nembach H, Fassbender J (2004) Picosecond Optical Control of the Magnetization in Exchange Biased NiFe/FeMn Bilayers. J Appl Phys 95(11):6613–6615. doi:10.1063/1.1652416
30. Longa FD, Kohlhepp JT Jonge, W.J.M.d., Koopmans, B. (2008): Laser-Induced Magnetization Dynamics in Co/IrMn Exchange Coupled Bilayers. J Appl Phys 103 (7):07B101. doi:10.1063/1.2830230
31. Dalla Longa F, Kohlhepp JT, de Jonge WJM, Koopmans B (2010) Resolving the Genuine Laser-Induced Ultrafast Dynamics of Exchange Interaction in Ferromagnet/Antiferromagnet Bilayers. Phys Rev B 81(9):094435. doi:10.1103/PhysRevB.81.094435
32. Mathieu C, Jorzick J, Frank A, Demokritov SO, Slavin AN, Hillebrands B, Bartenlian B, Chappert C, Decanini D, Rousseaux F, Cambril E (1998) Lateral Quantization of Spin Waves in Micron Size Magnetic Wires. Phys Rev Lett 81(18):3968–3971. doi:10.1103/PhysRevLett.81.3968
33. Jung S, Watkins B, DeLong L, Ketterson JB, Chandrasekhar V (2002) Ferromagnetic Resonance in Periodic Particle Arrays. Phys Rev B 66(13):132401. doi:10.1103/PhysRevB.66.132401
34. Kruglyak VV, Barman A, Hicken RJ, Childress JR, Katine JA (2005) Picosecond Magnetization Dynamics in Nanomagnets: Crossover to Nonuniform Precession. Phys Rev B 71:220409. doi:10.1103/PhysRevB.71.220409
35. Kruglyak VV, Barman A, Hicken RJ, Childress JR, Katine JA (2005) Precessional Dynamics in Micro-Arrays of Nanomagnets. J Appl Phys 97:10A706. doi:10.1063/1.1849057
36. Keatley PS, Kruglyak VV, Neudert A, Galaktionov EA, Hicken RJ, Childress JR, Katine JA (2008) Time-Resolved Investigation of Magnetization Dynamics of Arrays of Nonellipsoidal Nanomagnets with Nonuniform Ground States. Phys Rev B 78(21):214412. doi:10.1103/ PhysRevB.78.214412
37. Barman A, Wang S, Maas JD, Hawkins AR, Kwon S, Liddle A, Bokor J, Schmidt H (2006) Magneto-Optical Observation of Picosecond Dynamics of Single Nanomagnets. Nano Lett 6 (12):2939–2944. doi:10.1021/nl0623457
38. Barman A, Wang S, Maas J, Hawkins AR, Kwon S, Bokor J, Liddle A, Schmidt H (2007) Size Dependent Damping in Picosecond Dynamics of Single Nanomagnets. Appl Phys Lett 90(20):202504. doi:10.1063/1.2740588

39. Rana B, Pal S, Barman S, Fukuma Y, Otani Y, Barman A (2011) All-Optical Excitation and Detection of Picosecond Dynamics of Ordered Arrays of Nanomagnets with Varying Areal Density. Appl Phys Express 4(11):113003. doi:10.1143/APEX.4.113003

40. Rana B, Kumar D, Barman S, Pal S, Fukuma Y, Otani Y, Barman A (2011) Detection of Picosecond Magnetization Dynamics of 50 nm Magnetic Dots down to the Single Dot Regime. ACS Nano 5(12):9559–9565. doi:10.1021/nn202791g

41. Rana B, Barman A (2013) Magneto-Optical Measurements of Collective Spin Dynamics of Two-dimensional Arrays of Ferromagnetic Nanoelements. SPIN 03(01):1330001. doi:10.1142/s2010324713300016

42. Saha S, Mandal R, Barman S, Kumar D, Rana B, Fukuma Y, Sugimoto S, Otani Y, Barman A (2013) Tunable Magnonic Spectra in Two-Dimensional Magnonic Crystals with Variable Lattice Symmetry. Adv Func Mater 23(19):2378–2386. doi:10.1002/adfm.201202545

43. Mahato BK, Rana B, Kumar D, Barman S, Sugimoto S, Otani Y, Barman A (2014) Tunable Spin Wave Dynamics in Two-Dimensional $Ni_{80}Fe_{20}$ Nanodot Lattices by Varying Dot Shape. Appl Phys Lett 105(1):012406. doi:10.1063/1.4890088

44. Mahato BK, Rana B, Mandal R, Kumar D, Barman S, Fukuma Y, Otani Y, Barman A (2013) Configurational Anisotropic Spin Waves in Cross Shaped $Ni_{80}Fe_{20}$ Nanoelements. Appl Phys Lett 102(19):192402. doi:10.1063/1.4804990

45. Mahato BK, Choudhury S, Mandal R, Barman S, Otani Y, Barman A (2015) Tunable Configurational Anisotropy in Collective Magnetization Dynamics of $Ni_{80}Fe_{20}$ Nanodot Arrays with Varying Dot Shapes. J Appl Phys 117(21):213909. doi:10.1063/1.4921976

46. Barman A, Wang S, Hellwig O, Berger A, Fullerton EE, Schmidt H (2007) Ultrafast Magnetization Dynamics in High Perpendicular Anisotropy $[Co/Pt]_N$ Multilayers. J Appl Phys 101(9):09D102. doi:10.1063/1.2709502

47. Sajitha EP, Walowski J, Watanabe D, Mizukami S, Wu F, Naganuma H, Oogane M, Ando Y, Miyazaki T (2010) Magnetization Dynamics in CoFeB Buffered Perpendicularly Magnetized Co/Pd Multilayer. IEEE Trans Magn 46(6):2056–2059. doi:10.1109/tmag.2009.2038929

48. Pal S, Rana B, Hellwig O, Thomson T, Barman A (2011) Tunable Magnonic Frequency and Damping in $[Co/Pd]_8$ Multilayers. Appl Phys Lett 98(8):082501. doi:10.1063/1.3559222

49. Malinowski G, Longa FD, Rietjens JHH, Paluskar PV, Huijink R, Swagten HJM, Koopmans B (2008) Control of Speed and Efficiency of Ultrafast Demagnetization by Direct Transfer of Spin Angular Momentum. Nat Phys 4(11):855–858. doi:10.1038/nphys1092

Chapter 6
Electrical and Optical Control of Spin Dynamics

As discussed in previous chapters, in early days, the magnetization dynamics was primarily investigated using high-frequency techniques based on electrical excitation and detection. Janossy and Monod in 1976 [1] and Silsbee et al. in 1979 [2] found the signature of coupling between a dynamic ferromagnetic magnetization and spin accumulation in adjacent normal metals. It was demonstrated that a coating of ferromagnetic layer significantly enhances the microwave transmission through normal metal foils due to the assistance from conduction electron spin transfer. Subsequently, Grünberg et al. in 1986 [3] observed the non-local oscillatory exchange coupling, following which in 1988 Baibich et al. [4] and in 1989 Binasch et al. [5] reported the giant magnetoresistance (GMR) effect in their seminal works. Interestingly, in these studies, it was found that the electric current in a magnetic multilayer consisting of a sequence of thin magnetic layers separated by equally thin non-magnetic metallic layers is strongly influenced by the relative orientation of the magnetizations of the magnetic layers. It was noted that the resistance of the magnetic multilayer is low when the magnetizations of all the magnetic layers are parallel; however, this value becomes large when the magnetizations of the neighboring magnetic layers are ordered antiparallel. From this, it was inferred that during the transport of electric charge, the magnetic moment of electrons associated with the spin plays significantly important role. Although GMR may be observed in various geometries, applications typically employ ultrathin magnetic multilayers [6, 7]. This was followed by the discovery of tunneling magnetoresistance (TMR) in which ferromagnetic electrodes are separated by a thin insulating layer that serves as a tunneling barrier [8, 9]. A significant change in the tunneling conductance (usually an increase) is often observed when an applied magnetic field aligns the moments in the two ferromagnetic layers.

After the discovery of GMR in 1988, most of the concepts related to the study of spin dynamics required the generation of spin-polarized current [10–12]. Thus, a key to develop understanding of spin dynamics triggered by electrical means is to get an idea of spin polarization. The spin of an electron can be considered as a degree of freedom similar to charge which follows continuity equation and can be

© Springer International Publishing AG 2018

A. Barman and J. Sinha, *Spin Dynamics and Damping in Ferromagnetic Thin Films and Nanostructures*, https://doi.org/10.1007/978-3-319-66296-1_6

defined by the polarization vector. It is important to understand here the actual meaning of spin polarization of an ensemble of electrons. A system is said to be spin polarized if the two spin states have unequal distribution of population. In other words, spins have a preferential orientation which is analogous to the polarization of light where the electric field vectors are preferentially orientated along a particular direction [13]. An electron beam is called fully polarized if all the spins are in the same state. If the majority of spins belong to a particular state, then the beam is polarized to that state, and if they are equally distributed, then the beam is called unpolarized. The spin polarization of electron is important in many aspects. In theoretical studies, spin polarization makes it easier to calculate parameters like initial and final energy state of an electron system undergoing spin-dependent processes. Spin-polarized current can exert torque on magnetization, and thereby control magnetization dynamics and in turn magnetization reversal. In this chapter of the book, we will describe the magnetization dynamics with a perspective of understanding the finer details when the dynamics is excited by electric current or optical pulse.

6.1   Spin Dynamics Triggered by Electrical Current

In terms of applications, in spintronic devices, it is crucial to understand how the magnetization dynamics is triggered by electrical current. Most importantly, electrical current-induced spin manipulation has various advantages such as easy miniaturization and large-scale integration. Apart from applications, this understanding has led to unravel various fundamental mechanisms in the field of spin dynamics. In the field of electrical current-induced spin dynamics, year 1996 can be considered as a revolutionary year when Berger [14] and Slonczewski [15] independently predicted in their theoretical studies that the electric currents can cause a reorientation of the ferromagnetic order in multilayer structures. When an electron spin carried by the current interacts with a magnetic layer, the exchange interaction leads to torque between the spin and the magnetization. The torque that results from this interaction excites the magnetization when the current is large enough. The flow of spins is determined by the spin-dependent transport properties, like conductivity, interface resistance, and spin-flip scattering in the magnetic multilayer. Furthermore, it was shown that the transfer of vector spin momentum accompanying an electric current flowing through the interfaces of two magnetic films separated by a non-magnetic metallic spacer (magnetic double layer) can result in negative Gilbert-like (anti-damping) torque. If the current density is sufficiently high, then this torque leads to spontaneous magnetization precession as well as magnetization switching. These predictions led to the sophisticated design of sample structures and experiments to observe magnetization dynamics under the influence of current. Tsoi et al. in 1998 [16] experimentally demonstrated magnetization precession in $[Co/Cu]_N$ multilayers by currents injected by a point contact, whereas Myers et al. in 1999 [17] observed switching in the orientation of magnetic moments in Co/Cu/Co sandwich structures by perpendicular electrical currents. It may be noted that in spin-polarized current-induced switching, the roles of current and magnetic moments are reversed compared to GMR.

Apart from the GMR devices, current-induced spin dynamics have been actively studied in other magnetic nanostructures. Interestingly, in case of the domain wall, where the magnetization rotates continuously to connect two regions of uniform and antiparallel magnetization, a suitable injection of current pulse can lead to the displacement of the domain wall [18]. The current driven domain wall motion has huge application potential in the magnetic recording industry, and it has been actively investigated by various research groups [19, 20]. An interesting feature of domain wall motion is that the direction to which the domain wall moves depends on the electric current polarity. Also, triggering of magnetization dynamics in ferromagnetic thin-film heterostructures has also been pursued by various research groups [21–25]. The combination of charge and spin transport in heterostructures has been emerging as an intriguing field in last few years. Given the large volume of interesting work in the field of electrical current-induced magnetization dynamics, we try to focus on a small part of experimental and theoretical studies which, we believe, will enable readers to get flavor of this interesting topic.

6.2 Spin-Transfer Torque

In order to understand of spin-transfer torque, let us consider an important magnetic nanostructure, namely, spin valve, where a thin-film non-magnet (spacer) is sandwiched between two thin-film ferromagnets [10, 26]. We assume that in the range of film thicknesses most commonly used, the magnetization M of the 'fixed' layer and the magnetization m of the thinner 'free' layer lie in the plane of the film. M and m form an angle θ (Fig. 6.1). When a charge current flows through a ferromagnet, it becomes spin polarized. The spin-polarized current carries angular momentum, and the current remains polarized in neighboring non-magnetic layers. Interestingly, the angular momentum carried by the current interacts with the magnetization of the magnetic layers and exerts a spin-transfer torque on the magnetization [14, 15]. The electrons are injected perpendicularly to the plane of the layer. The direction of the spin polarization in the normal metal cannot be parallel to both M and m. Thus, when the two magnetizations are non-collinear, the spin polarization direction makes an angle with respect to M. When the electrons are passing through M, they align their spins by the exchange interaction in the direction of m. Since the exchange interaction is spin conserving, this means that the transverse component of the spin current has been absorbed and transferred to m. This transfer of spin angular momentum is equivalent to the existence of a STT acting on m. In the framework of Slonczewski [15, 27], the expression of the time

Fig. 6.1 Schematic diagram of the spin valve structure. The device is made of two magnetic layers with magnetization M and m separated by a non-magnetic layer. The *bottom* layer is thick enough to be considered as a spin polarizer of the current and fixed upon the action of the current. The *top* layer is thin and free to move under the action of the spin transfer torque. The magnet transmits and scatters the collinear component of the spin $s_{//}$ and absorbs the transverse component $s_\perp$. This results in a torque that can excite or reverse the magnetization m

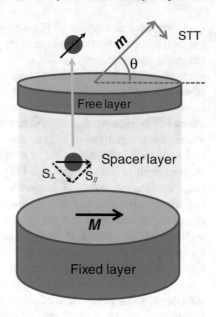

variation of the magnetization m of the thin ferromagnetic layer due to the spin transfer is as follows:

$$\frac{\mathrm{d}m}{\mathrm{d}t} = -P_{\mathrm{trans}}m \times (m \times M) \tag{6.1}$$

where the parameter P_{trans} is proportional to the magnitude of the transverse component of the spin-polarized current and is proportional to the current (I) and the spin polarization ($\eta(\theta)$). P_{trans} is usually expressed as

$$P_{\mathrm{trans}} = \eta(\theta)\frac{I\hbar}{2e}\hat{M} \tag{6.2}$$

and thus, the modified LLG equation including STT term can be represented as the following equation:

$$\frac{\mathrm{d}m}{\mathrm{d}t} = -\gamma\left(m \times H_{eff}\right) + \alpha\left(m \times \frac{\mathrm{d}m}{\mathrm{d}t}\right) - \eta(\theta)\frac{I\hbar}{2e}m \times (m \times M) \tag{6.3}$$

It is important to mention here that the torque due to the transfer of spin has the same form as the second term, i.e. the damping torque. Thus, depending on the sign of the current it can either dissipate or provide energy to the system. Typically, in metallic ferromagnets, the value of the critical current densities required to observe the effect of STT, excitation, and/or reversal of the magnetization is about 10^7 A/cm^2 [28].

Various research groups have successfully demonstrated the spin-transfer-induced magnetization dynamics since the prediction of the effect in 1996. A number of device geometries such as point contact [16], lithographically defined nanopillars [28, 29], electrochemically grown nanowires [30], and magnetic tunnel junctions [31] have been investigated experimentally to observe effect of STT. Two important requirements to observe STT in these devices are small cross-sectional area and magnetoresistive readout of the magnetic state. The strict requirement for small cross-sectional areas in the devices is due to two main reasons [26]. First is related to the relative effect of the Oersted field which decreases with the decrease in cross-sectional area. Thus, in order to observe the spin-transfer torque effect, it is important to reduce the effect of the Oersted field which is generated by the current passing through the system. Large self-field can influence the direction of the magnetic moments and may complicate the spin dynamics as well as reversal process. Second reason is related to high current density which is required to transfer sufficient angular momentum to affect the magnetization. The heat generated by the current may be detrimental for the devices if it is not confined to significantly small area which is in good thermal contact to a large mass. In the devices where STT is observed, the readout is mainly performed using the GMR effect, as the resistance of the device depends on its magnetic state [10]. It specifically allows the magnetic state to be inferred in the devices by measuring its resistance as a function of current.

In general, it is preferred to keep fixed one of the layers by choosing it as thicker film or by pinning it using exchange bias effect (by coupling it to an antiferromagnet) so that it is less sensitive to spin-transfer torque. Typically, the signature of a spin-transfer effect is observed as a change in the resistance of the device which is asymmetric in the current. In Fig. 6.2, we reproduce the nanopillar device schematic from Katine et al. [28]. The nanopillar in their experiment was lithographically patterned, and it had dimension of 130 ± 30 nm. The multilayer film stack was sputter deposited on oxidized Si substrate with layers 120 nm Cu/10 nm Co/6 nm Cu/25 nm Co/15 nm Cu/3 nm Pt/60 nm Au. The variation of differential

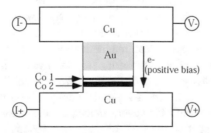

Fig. 6.2 Schematic of pillar device with Co (dark) layers separated by a 60 Å Cu (light) layer. At positive bias, electrons flow from the thin (1) to the thick (2) Co layer. *Reprinted with permission from Ref. [28]. Copyright 2000 by the American Physical Society*

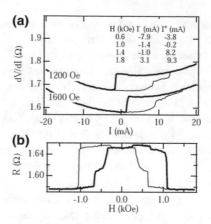

Fig. 6.3 a *dV/dI* of a pillar device exhibits hysteretic jumps as the current is swept. The current sweeps begin at zero; light and dark lines indicate increasing and decreasing current, respectively. The traces lie on top of one another at high bias, so the 1200 Oe trace has been offset vertically. The inset table lists the critical currents at which the device begins to depart from the fully parallel configuration (I^+) and begins to return to the fully aligned state (I). **b** Zero-bias magnetoresistive hysteresis loop for the same sample. *Reprinted with permission from Ref.* [28]. *Copyright 2000 by the American Physical Society*

resistance of the nanopillar as a function of applied current is shown in Fig. 6.3a, whereas the conventional resistance versus field plot at zero current is shown in Fig. 6.3b. From Fig. 6.3a, it can be noticed that the curve is hysteretic, with the device remaining in this high-resistance state until the current is swept to negative values where it returns to the low-resistance orientation in a single jump. Additionally, the resistance gradually rises with further increase in the current bias due to the growth of electron-magnon and electron-phonon scattering. The hysteretic jump in resistance is due to the relative alignment of the magnetization of the Co layers, with the low-resistance state reflecting the parallel alignment of the layer magnetizations, and the high-resistance state corresponding to antiparallel alignment. Interestingly, the magnitudes of the change in resistance versus current and resistance versus field curve were found to be the same.

Though in the above-discussed experiments, STT-induced switching of magnetization in a bistable magnetic devices was established, more sophisticated experiments later showed even clean magnetization, clean switching between fully parallel and fully antiparallel alignments (square hysteresis loops) [32]. It was understood that the fields greater than the coercive field inhibit a stable antiparallel state, and the system switches from a parallel configuration into more complicated states including precession. However, detecting the magnetization precession required the measurement of the time or frequency response of the resistance. In 2003, Kiselev et al. [29] first reported the precessing state of magnetization in current driven dynamics experiments using STT through peaks in the power spectrum density in frequency-dependent measurements which opened the new device concept of spin torque nano-oscillators. In their experiment, the sputter-deposited multilayer of composition 80 nm Cu/40 nm Co/10 nm Cu/3 nm Co/2 nm Cu/30 nm Pt onto an oxidized silicon wafer was patterned to an elliptical nanopillar with cross-sectional area 130 nm × 70 nm. The passage of dc current through the thicker layer of the device creates spin-polarized current which can apply a torque to the thinner Co layer. Continuous application of the current can subsequently set oscillations of the free-layer magnetization relative to the fixed layer thereby continuously changing the device resistance. The magnetization dynamics triggered by this mechanism produces a time-varying voltage with typical frequencies in the microwave range.

In Fig. 6.4a–d, we reproduce the results obtained by Keislev et al. in their experimental investigations [29]. Figure 6.4a shows the nanopillar schematic along with the heterodyne mixer circuit for measurement of high-frequency microwaves. In Fig. 6.4b, differential resistance versus current for STT driven hysteretic switching between low-resistance parallel and high-resistance antiparallel states are shown. For large applied field, the current produces peaks in the differential resistance which is known to be associated with dynamical magnetic excitation. Inset of Fig. 6.4b shows the magnetization switching when the field is swept. In Fig. 6.4c, d, the microwave spectrum measured at different applied dc current is shown. Interestingly, the peaks in the spectrum corresponded to small-angle elliptical precession of the free layer, thereby confirming pioneering predictions that spin transfer can coherently excite the uniform spin-wave mode in the nanomagnets. Following this work, a number of experiments have tested the predictions of Slonczewski's model for spin-transfer torques [33–37]. Furthermore, various studies have compared predictions of macrospin simulations with experimental results [26, 38]. Interestingly, the theoretically computed phase diagrams of microwave power as a function of applied magnetic field and current have been found to qualitatively agree in many cases with those measured experimentally.

Note that from the above-described results, it is evident that the precessional motion of the magnetization in a suitable device can convert a DC input current into an AC output voltage. Such behavior might be useful for making current-controlled oscillators. One of the desired properties of an oscillator is to obtain large enough output power in the microwave emission along with high Q-factor [39]. It is worth to mention here that several dedicated studies with an aim to achieve large output power, high-microwave frequency, and significantly narrow linewidth have been performed in recent past [35, 36, 40, 41]. Different device geometry (point contact, etc.,), various ferromagnetic materials including large spin-polarization Heusler alloys and various spacer layers have been tested to achieve the above-mentioned goal, and it is still an active area of ongoing research in the spintronics community [36, 42].

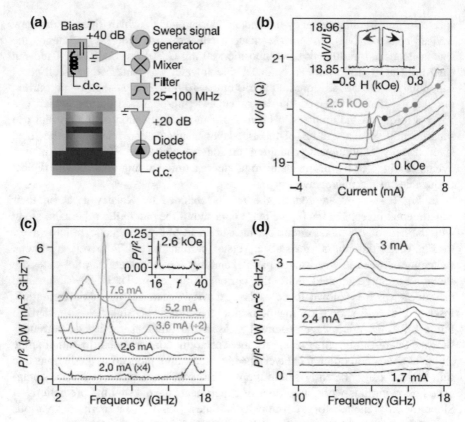

Fig. 6.4 Resistance and microwave data for sample. **a** Schematic of the sample with copper layers (orange), cobalt (blue), platinum (green) and SiO_2 insulator (grey), together with the heterodyne mixer circuit. Different preamplifiers and mixers allow measurements over 0.5–18 GHz or 18–40 GHz. **b** Differential resistance versus current for various magnetic fields of 0 (*bottom*), 0.5, 1.0, 1.5, 2.0, and 2.5 kOe (top), with current sweeps in both directions. At $H = 0$, the switching currents are $I'_c = 0.88$ mA and $I_c = -0.71$ mA, and $\Delta R_{max} = 0.11$ Ω between the P and AP states. Coloured dots on the 2 kOe curve correspond to spectra shown in (c). Inset to (b), Magnetoresistance near $I = 0$. Red and black indicate different directions of magnetic-field sweep. **c** Microwave spectra (with Johnson noise subtracted) for $H = 2.0$ kOe, for $I = 2$mA (bottom), 2.6, 3.6, 5.2 and 7.6 mA (top). Power density divided by I^2 is plotted to facilitate comparisons of the underlying changes in resistance at different current values. Inset to (c), spectrum at $H = 2.6$ kOe and $I = 2.2$ mA, for which both f and $2f$ peaks are visible on the same scan. **d** microwave spectra at $H = 2.0$ kOe, for current values from 1.7 to 3.0 mA in 0.1 mA steps, showing the growth of the small-amplitude precessional peak and then a transition in which the second harmonic signal of the large-amplitude regime appears. *Reprinted with permission from Ref.* [29]. *Copyright 2003 by the Nature Publishing Group*

6.3   Current-Induced Spin–Orbit Torque

Another way of electrical control of the magnetization dynamics is by using a recently discovered mechanism namely spin–orbit torque (SOT) [43]. The SOT can efficiently excite or reverse the magnetization direction and is therefore potentially useful for applications such as magnetic memories or logic. The SOT exploits both the flow of spin-polarized current and static electric fields [44]. Importantly, SOT explicitly depends on the presence of strong SO coupling intrinsic to the nuclear composition and atomic structure of a material, thus it is fundamentally different from STT. In case of non-magnetic semiconductors, D'yakonov and Perel in 1971 predicted correlation between charge current and spin polarization arising from the spin–orbit interaction (SOI) [45]. After several decades, in 2009, Manchon and Zhang in a theoretical study showed that in a single ferromagnetic layer as well, the SOI may give rise to a transverse spin density which exerts a torque on the local magnetization [46]. Subsequently, two separate pioneering experimental studies by Miron et al. in 2010 and 2011 showing the evidence of SOT in Pt/Co/AlO$_x$ heterostructures opened a new paradigm in the field of modern spintronics research involving metallic ferromagnets [21, 22]. In one of the works, they observed the switching of a perpendicularly magnetized cobalt dot driven by in-plane current injection without the need of any polarizing layer, and in another work, they reported extremely fast domain-wall motion in nanowires made of similar heterostructures. They explained these results based on the SOTs, primarily originating from Rashba effect which takes into account the effective magnetic fields generated from the combined action of an electric current and an asymmetric crystal field intrinsic to materials lacking inversion symmetry [47]. In a separate study in 2012, Liu et al. [23] showed magnetization switching in Ta/CoFeB/MgO heterostructures by considering the spin torque originating from the spin current generated due to spin Hall effect (SHE) in the tantalum layer. Following the above-mentioned studies, in last few years, control of spin dynamics has been experimentally established by transferring spin angular momentum between a flowing spin current and the local magnetization of ferromagnet by alternative mechanisms (SOT) as compared to STT conventionally observed in magnetic spin valve structures and magnetic domain walls; however, the understanding still remains elusive [43, 48, 49]. It is being commonly considered that there are two mechanisms for the origin of SOT: (1) Rashba–Edelstien interfacial effect [47, 50] and the (2) bulk SHE [45, 51]. While the contribution of SHE on SOT has been studied by several groups, there have been only few studies focusing on the interfacial contribution to SOT. An extremely challenging issue up to till date is to identify the physical origin of the SOT, as the SHE and Rashba-like effect are known to produce effects those are intricately related [43, 49]. Considering the situation that this research area is rapidly evolving, we provide a glimpse of interesting results in this field along with brief qualitative explanation of Rashba effect and SHE.

In 1984, Bychkov and Rashba investigated the case of an ordinary two-dimensional electron gas (2DEG) in the absence of a mirror plane either due to perpendicular electric field or asymmetry in bonding such as in a bilayer heterostructure [47]. They proposed that in 2DEG the Hamiltonian contains a momentum-dependent spin–orbit field of the Rashba form $B_{\text{Rashba}} \propto \alpha_R(k \times z)$ where α_R is a phenomenological coefficient related to the spin–orbit coupling constant λ, k is the electron momentum, and z is the sample-normal direction. Interestingly, the Rashba field shifts the energies of electron states up or down depending on whether their spins are parallel or antiparallel to B_{Rashba}, i.e. it lifts the spin degeneracy of the surface or interface state Fermi surfaces. Because of time-reversal symmetry, the spin orientation of a surface band state depends on momentum, and on a given Fermi surface, it is opposite in case of opposite momentum. Because an in-plane current increases the occupation probability of states on one side of the Fermi surface and decreases them on the other side, it naturally generates a non-equilibrium spin accumulation. If this non-equilibrium spin accumulation is exchange coupled to an adjacent magnetic layer, it can apply a significant torque to influence the spin dynamics [46, 52].

In 1999, Hirsch in a theoretical study predicted interesting phenomena for paramagnetic metals. In metals with strong spin–orbit coupling (either intrinsically or from impurities), flow of charge current can lead to spin currents that flow perpendicular to the direction of charge current and produce spin accumulations at the sample boundaries, and the phenomena was referred as SHE [53, 54]. Most importantly, this spin accumulation can result in a spin current into a neighboring layer, without accompanying charge current. Thus, if a ferromagnetic layer is placed as a neighboring layer, then it may experience a spin torque due to the spin current generated by SHE. The mechanism of SHE has been described earlier in terms of intrinsic spin–orbit coupling in the heavy metal layer, side jump, and skew scattering from impurity center in the non-magnetic metal. Also, inverse of SHE is known to exist in which a pure spin current leads to charge accumulation and development of finite voltage in a direction transverse to the spin current [55, 56]. The SHE is quantified in terms of material-specific spin Hall angle which is given by the ratio of charge-to-spin current densities. This parameter describes the conversion efficiency from charge current to spin current. It is worth to mention here that while the SHE depends on spin–orbit coupling, its calculation requires detailed band structure knowledge in both the intrinsic and extrinsic (impurity/disorder) limit. The SHE in transition metal elements undergoes sign reversal with increasing band filling across the periodic table [57], and large value of spin Hall angle is reported for non-magnetic heavy metals such as Pt, Ta, or W with large spin–orbit coupling [58]. It is more or less understood that the bulk SHE directly affects the switching behaviors of a ferromagnetic layer. By changing the bulk SHE of a material, one can increase the magnetization switching efficiency, thus decreasing critical current required to switch the magnetization in magnetic tunnel junction used in magnetic random access memories. In fact, there are several reports on the contribution of bulk SHE on SOTs and many researchers attempted to manipulate bulk SHE.

In 2013, Haney et al. using semiclassical calculations showed that the torques due to the SHE and Rashba–Edelstein effect can be treated simultaneously [59]. In the semiclassical approximation, the interfacial SOI acting on the spins passing through the two-dimensional interface creates a spin polarization. Thus, at the interface, spin polarization gets coupled to the magnetization via exchange coupling. The form of the torque observed for the case of interface spin–orbit coupling is found to be exactly the same as those due to the SHE with different efficiency coefficient. For the torques due to interfacial spin–orbit coupling, the field-like contribution is expected to be larger than the damping-like contribution. Following first principle calculations as well other groups have obtained similar results with both damping-like and field-like torques having contributions both from the bulk and the interface [60, 61]. To convincingly separate out, the contribution from SHE and Rashba effect in experiments is quite challenging as one can only measure the symmetry of the torque, and the SHE may not be the only contribution with that particular symmetry. The experimental realization of SOT is performed by estimating the current-induced effective field along the direction of flow of current (longitudinal effective field) and perpendicular to the current (transverse effective field) [49].

The general form of the modified Landau–Lifshitz–Gilbert equation that takes into account STT terms together with the SOT in a magnetic thin-film heterostructure may be expressed as follows:

$$\frac{d\boldsymbol{m}}{dt} = -\gamma\left(\boldsymbol{m}\times\boldsymbol{H}_{eff}\right) + \alpha\left(\boldsymbol{m}\times\frac{d\boldsymbol{m}}{dt}\right) - a_J\boldsymbol{m}\times(\boldsymbol{m}\times\boldsymbol{p}) - \gamma b_J(\boldsymbol{m}\times\boldsymbol{p}) \quad (6.4)$$

where $\boldsymbol{p}$ is the direction of spin polarization, the third term is the damping-like term, and the fourth term is the field-like term [62–64].

Various experimental studies have been performed by several groups to obtain in-depth understanding of spin dynamics governed by SOT. Liu et al. in 2012 reported an interesting study in which using the spin torque from SHE in tantalum microstrip, persistent steady state magnetic oscillations in an adjacent nanomagnet was induced [48]. The samples were made from a $Ta(6)/Co_{40}Fe_{40}B_{20}(1.5)/MgO(1.2)/Co_{40}Fe_{40}B_{20}(4)/Ta(5)/Ru(5)$ thin-film stack (thicknesses in nanometers). The stack was first patterned into a microstrip that is 1.2 μm wide and 6 μm long. The CoFeB/MgO/CoFeB magnetic tunnel junction was then milled into a 50×180 nm^2 nanopillar on the Ta microstrip. In Fig. 6.5a, the schematic diagram of sample with SHE is shown; when a dc current is applied along the Ta strip the SHE injects spin current into the CoFeB free layer. If the injected spin moments are antiparallel to the free layer, they will exert spin torque. In this experiment, the magnetic oscillations were observed using a current-biased magnetic tunnel junction in contact with the oscillating element to produce a microwave output voltage across the MTJ. Interestingly, the microwave emission spectrum (cf. Fig. 6.5b) was found only for the negative direction of the flow of current through the Ta layer, and the spin current through Ta was inferred to be dominant for exciting magnetic oscillation. Furthermore, this study suggested an independent control of frequency and amplitude of microwave spectrum using spin torque from SHE in spin torque nano-oscillators.

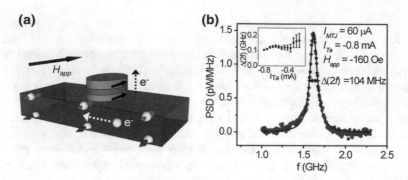

Fig. 6.5 **a** Schematic showing the direction of spin accumulation induced by the SHE at the top and bottom surface of Ta strip by a positive current I_{Ta}. The dashed arrows denote the directions of electron flow for positive applied currents I_{Ta} and I_{MTJ}. H_{app} shown in the figure corresponds to a negative field, the direction used for measurements of microwave spectra. **b** Microwave spectrum for $I_{Ta} = -0.8$ mA with a fit to a Lorentzian peak. The linewidth of the second harmonic peak is 104 MHz. Inset: dependence of the linewidths for the second harmonic peaks as a function of I_{Ta}. *Reprinted with permission from Ref.* [48]. *Copyright 2012 by the American Physical Society*

6.4   Optical Control of Spin Dynamics

Spin dynamics directly induced by optical pulse has been described in Chap. 2. The experimental means of measurement of optically induced spin dynamics have been described in detail in Chap. 4. In addition to direct excitation of spin dynamics by femtosecond laser pulse, it has also been excited by producing a magnetic field pulse using a photoconductive switch and a transmission line/coplanar waveguide structure in 1990s [65–67]. The ensuing magnetization dynamics were measured by using a time-resolved scanning Kerr microscope or a simple time-resolved Kerr magnetometer based on a pump-probe technique as described in the literature [68, 69]. Over the years, a number of different techniques and physical mechanisms have been used for the coherent control of spin dynamics, and here, we will focus mainly on the optical coherent control, where a laser pulse is used to indirectly or directly control the spin precession in magnetic thin films and patterned elements.

6.4.1   Coherent Control of Precessional Dynamics by Magnetic Field Pulse Shaping

The ever increasing demand of faster switching in smaller data bit calls for precessional switching, which is potentially few orders of magnitude faster than the conventional switching processes. However, continuation of precessional oscillation of magnetization vector after switching, also known as ringing, puts a limitation on the switching time. Heavily damped materials may reduce the ringing time but cannot completely eliminate it immediately after switching. Hence, coherent control of magnetization precession by shaping the magnetic field pulse has attracted the magnetics community from early 2000. It is believed that the use of coherent control methods will allow the operation of magnetic storage and memory devices at fundamental frequency limits by eliminating the adverse effects of precessional oscillations. Efforts have been made on achieving coherent control on precessional oscillation in ferromagnetic thin films by controlling the magnetic field pulse generated electronically [70–73]. Later, Schumacher et al. achieved coherent control of precessional dynamics and showed evidence of multiple coherent periodic precessional switching in microscopic spin valve elements by ultrashort field pulses as short as 140 ps [74, 75]. In 2002, Gerrits and co-workers have addressed the issue of ringing after precessional switching by engineering the field pulse profile. In this double-pulse scheme, two separate GaAs photoswitches were triggered independently by femtosecond laser pulse excitation [76]. Desired temporal profile of the field pulse was obtained by superimposing these two pulses (Fig. 6.6). Pump pulse 1 was used for initiating the precession, and the pump pulse 2 acted as a stop pulse. The time delay between these start and stop pulses was precisely varied by using a delay line. Coherent suppression of the ringing was observed when this delay time was exactly half the period of the precession. The reversal time was

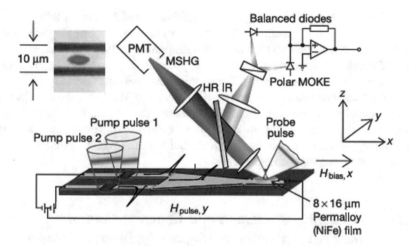

Fig. 6.6 Schematic of experimental setup. A sequence of two pump laser pulses, which are separated by a distinct time delay, excites two GaAs-photoconductive switches. The generated current pulses are superimposed to produce one short and square-like magnetic field pulse. The field pulse is launched down a coplanar waveguide structure and excites the thin-film permalloy ($Ni_{81}Fe_{19}$) element at the end of the tapering. The *Inset* is a microscopic photograph of the 8-nm-thin permalloy magnetic element that has an elliptical shape with dimensions of $8 \times 16 \ \mu m^2$. The vector- and time-resolved element response is measured by magnetization-induced second harmonic generation (MSHG) and the polar magneto-optical Kerr effect (MOKE). A high-reflectance infrared mirror (HRIR) is used to split the fundamental and second harmonic part of the beam. A photomultiplier tube (PMT) is used to detect the second harmonic photons. *Reprinted with permission from Ref. [76]. Copyright 2002 by the Nature Publishing Group*

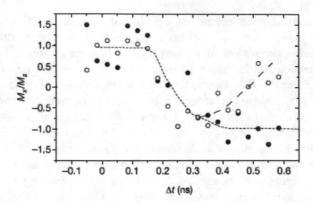

Fig. 6.7 Switching by large-field excitation and suppression of ringing. Without a stop pulse, the system switches back to its initial state (*open circles*). After sending the stop pulse, the suppression of the ringing of the magnetization can clearly be observed (*solid circles*). The *lines* are guides to the eye. The low signal-to-noise ratio in the M_x component results from the very weak longitudinal MSHG signal with an incoming polarization parallel and second harmonic polarization perpendicular to the plane of incidence (p_{in}–s_{out}). *Reprinted with permission from Ref. [76]. Copyright 2002 by the Nature Publishing Group*

observed to be 200 ps, which suggests a switching rate of 5 GHz for this particular experiment. In Fig. 6.7, the authors had shown a suppression of ringing with a stop pulse in contrast to the case where no stop was applied. In 2005, Barman et al. [77] studied the spatial coherence of coherent suppression of precessional switching in circular microscale elements with 22 nm thickness and diameters of 10, 7, and 5 μm by using time-resolved scanning Kerr imaging. While the larger elements showed spatially uniform coherent suppression, the smallest element (5 μm) showed spatial non-uniformity in coherent suppression suggesting that a simple pulse shaping scheme will not be sufficient, and a more involved pulse shaping is required in confined magnetic elements.

6.4.2   Non-Thermal Excitation and Coherent Control of the Spin Dynamics in Magnets by Inverse Faraday Effect

The laser-induced ultrafast demagnetization and optically excited coherent spin waves are results of optical absorption followed by a rapid increase in the electron, spin and lattice temperatures. The possibility of complete non-thermal or opto-magnetic excitation and control of magnetization dynamics was first demonstrated by Kimel et al. in $DyFeO_3$, in the hundreds of gigahertz frequency range [78]. The optomagnetic excitation can be understood as follows [13, 79]. The thermodynamic potential of an isotropic, non-absorbing, magnetically ordered medium with static magnetization $M(0)$, and magneto-optic susceptibility α_{ijk} in a monochromatic light field $E(\omega)$ includes the term

$$F = \alpha_{ijk}E_i(\omega)E_j(\omega)^*M_k(0) \tag{6.5}$$

Consequently, the electric field of light at frequency ω will act on the magnetization as an effective magnetic field directed along the wave vector of the light $\mathbf{k}$ given by:

$$H_k = -\frac{\partial F}{\partial M_k} = -\alpha_{ijk}E_i(\omega)E_j(\omega)^* \tag{6.6}$$

For isotropic media, α_{ijk} is a fully antisymmetric tensor with a single independent element α, and we may rewrite Eq. (6.6) as follows:

$$\mathbf{H} = \alpha[\mathbf{E}(\omega) \times \mathbf{E}(\omega)^*] \tag{6.7}$$

which means right- and left-circularly polarized lights should act as magnetic fields of opposite sign. Hence, in addition to the magneto-optical Faraday effect where the plane of polarization of a linearly polarized light gets rotated upon transmission through a magnetic medium, there can also be an inverse Faraday effect caused by the same susceptibility α. One may note that as opposed to conventional magnetic resonance, the electric field component of the light is responsible for inverse Faraday effect.

In DyFeO$_3$, the antiferromagnetically coupled Fe spins are slightly canted due to the Dzyaloshinskii–Moriya interaction, resulting a small spontaneous magnetization $M_S \sim 8G$, while it has a giant Faraday rotation of $3000°$ cm^{-1}. These enabled the detection of optically induced magnetization by measuring direct magneto-optical Faraday effect. Figure 6.8 shows the time-resolved Faraday rotation in a z-cut DyFeO$_3$ sample for two circularly polarized pump pulses of opposite helicities. The transient magnetic signal is followed by magnetization oscillation with a frequency of about 200 GHz, while the sign of the photo-induced magnetic signal depends on the helicity of the pump pulse. The latter unambiguously demonstrates the direct coupling between photons and spins in DyFeO$_3$. The magnetization oscillation is superposed on an exponential decay with time constant of about 100 ps, which corresponds to photo-induced change in the equilibrium orientation of the magnetization and a subsequent decay of this equilibrium orientation to the initial state. Further, experiment showed a linear dependence of the photo-induced spin oscillation on the pump intensity, which indicates the photo excitation of magnons is a two-photon process as also predicted by Eq. 6.7.

The non-thermal effect of light on magnetization inspires coherent control of spin precession by using two pump pulses separated by a controllable optical delay (Fig. 6.9). In this experiment, a pump pulses of helicity σ^+ is used to trigger

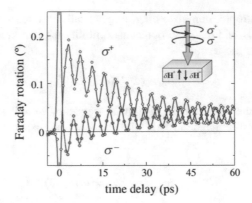

Fig. 6.8 Magnetic excitations in $DyFeO_3$ probed by the magneto-optical Faraday effect. Two processes can be distinguished: (1) instantaneous changes of the Faraday effect due to the photoexcitation of Fe ions and relaxation back to the high spin ground state $S = 5/2$; (2) oscillations of the Fe spins around their equilibrium direction with an approximately 5 ps period. The circularly polarized pumps of opposite helicities excite oscillations of opposite phase. *Inset* shows the geometry of the experiment. Vectors δH^+ and δH^- represent the effective magnetic fields induced by right-handed σ^+ and left-handed σ^- circularly polarized pumps, respectively. *Reprinted with permission from Ref.* [76]. *Copyright 2002 by the Nature Publishing Group*

antiferromagnetic spin precession in $DyFeO_3$ at the zero delay ($t = 0$), while a second pump of helicity σ^+ arrives after an odd number of half precessional period and shifts the magnetization further away from the effective magnetic field causing the precession to have about twice the amplitude of the original precession (amplification, Fig. 6.9b). In contrary, the second pump pulse arrives the sample after an integer number of full periods the magnetization comes back to its original equilibrium position and stops the precession (stopping, Fig. 6.9c). This clearly demonstrates that femtosecond optical pulse can be directly used for the coherent control of spin dynamics. The relative phase between the spin precession and the second pump pulse determines whether energy transfer will occur from the laser pulse to the magnetic system (amplification) or from the magnetic system to the laser pulse (stopping of precession).

Inspired by the fact that the circularly polarized femtosecond laser pulses act as equally short magnetic field pulses via the inverse Faraday effect, efforts have been made to completely reverse the magnetization of a magnetic domain by using this optically induced field pulses [80]. For this experiment, GdFeCo thin films are placed under a polarizing microscope where domains with magnetization 'up' and 'down' could be observed as white and black regions, respectively. Amplified pulses from a Ti: sapphire laser at a wavelength of 800 nm, repetition rate of 1 kHz, and a pulse width of 40 fs were used to excite the sample. The laser pulses were incident normal to the sample surface, so that an effective optically generated magnetic field would be directed along the magnetization, similar to a conventional recording scheme. The laser beam was swept at high speed across the sample, so that each pulse landed at a different spot.

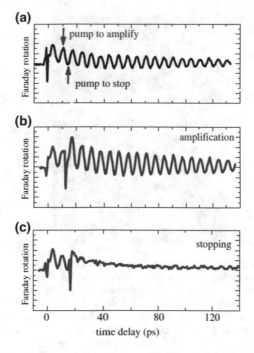

Fig. 6.9 Coherent control of spins in DyFeO₃ with two circularly polarized laser pulses at $T = 95$ K. **a** Precession triggered by the first laser pulse; **b** amplification of spin precession by the second laser pulse that comes after an even number of full periods; and **c** stopping of the spin oscillations by the second pump that comes after an odd number of half periods. (*Reprinted with permission from Ref.* [13]. *Copyright 2010 by the American Physical Society*

Figure 6.10a shows that σ^+ pulses reverse the magnetization in the black domain but does not affect the magnetization of the white domain. The opposite situation is observed for σ^- pulses, which reverse the white domain without affecting the black domain. Thus, during the presence of a single 40 fs laser pulse, information about the angular momentum of the photons is transferred to the magnetic medium, and subsequently, recording occurs. These experiments unambiguously demonstrate that all-optical magnetization reversal can be achieved by single ultrashort circularly polarized laser pulses without the aid of an external magnetic field. This is more important as a precessional switching within 40 fs would require effective magnetic fields above 100 T and unrealistically strong damping, while, such strong and short magnetic field pulses are not expected to lead to a deterministic switching of magnetic domains.

This experiment led to demonstration of optomagnetic recording as illustrated in Fig. 6.10b. It shows how optically written bits can be overlapped and made much smaller than the beam waist by modulating the polarization between σ^+ and σ^- as the laser beam is swept across the sample. High-density recording may also be achieved by employing near-field antenna structures similar to those used and

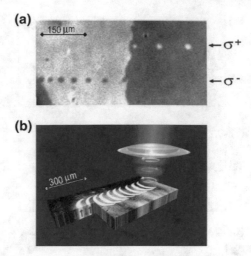

Fig. 6.10 All-optical magnetic recording by femtosecond laser pulses. **a** The effect of single 40 fs circularly polarized laser pulses on the magnetic domains in $Gd_{22}Fe_{74.6}Co_{3.4}$. The domain pattern was obtained by sweeping at high speed (50 mm/s) circularly polarized beams across the surface so that every single laser pulse landed at a different spot. The laser fluence was about 2.9 mJ/cm². The small size variation of the written domains is caused by the pulse-to-pulse fluctuation of the laser intensity. **b** Demonstration of compact all-optical recording of magnetic bits. This was achieved by scanning a circularly polarized laser beam across the sample and simultaneously modulating the polarization of the beam between *left* and *right* circular. *Reprinted with permission from Ref.* [80]. *Copyright 2007 by the American Physical Society*

further being developed for heat assisted magnetic recording. These experiments combined with emerging development of ultrafast lasers and nanotechnology may lead to the realization of a new generation of magnetic recording devices.

6.4.3 All-Optical Control of Ferromagnetic Thin Films and Nanostructures

The development of all-optical helicity-dependent switching (AO-HDS) was initially limited to ferrimagnetic systems such as rare earth–transition metal (RE–TM) alloys and synthetic ferrimagnets where two distinct sublattices are antiferromagnetically exchange coupled. The possible reasons behind AO-HDS were thought to be based on the existence of an effective field created by the circularly polarized light via the inverse Faraday effect or by the transfer of angular momentum from the light to the magnetic system. In 2011, Radu et al. claimed the formation of a transient ferromagnetic state due to different demagnetization times for RE and TM sublattices as the reason behind AO-HDS, where the light's helicity plays a secondary role [81]. Laser-induced superdiffusive spin currents may also flow in heterogeneous systems, potentially contributing to the AO-HDS process [82–86].

However, it is technologically imperative that AO-HDS does not remain limited to a subset of ferrimagnetic materials only but can also be applied in ferromagnetic materials, in particular, high-anisotropy granular or patterned materials that are anticipated for future high-density magnetic recording. Very recently, Mangin et al. [87] showed AO-HDS in a much broader variety of materials, including rare earth–transition metal alloys, multilayers, and heterostructures. They further showed that rare earth-free Co-Ir-based synthetic ferrimagnetic heterostructures designed to mimic the magnetic properties of rare earth–transition metal alloys also exhibit AO-HDS. In Fig. 6.11, we reproduce the magneto-optical image as obtained by Mangin et al. for Co/Tb multilayers [87]. Here, thermal demagnetization (TD) is characterized by the formation of magnetic domains with random up or down orientation that forms independent of the laser beam helicity as shown in Fig. 6.11a. In case of AO-HDS, interaction of laser with sample causes deterministic magnetization reversal depending on the helicity of the laser as shown in Fig. 6.11b.

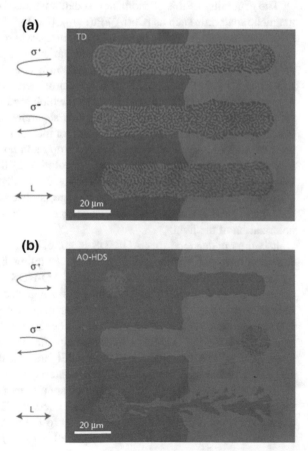

Fig. 6.11 Examples of the two optical responses for two different samples. **a, b** [Co (0.8 nm)/Tb(0.4 nm)]$_{\times 21}$ multilayers showing thermal demagnetization (TD) (**a**) and [Co(0.5 nm)/Tb(0.4 nm)]$_{\times 28}$ multilayers showing AO-HDS (**b**). For each sample, three types of polarized laser beam were swept over the sample, and the magnetization pattern was subsequently imaged: from *top* to *bottom*, *right*-circularly polarized light (σ^+), *left*-circularly polarized light (σ^-), and linearly polarized light (L). In the images, *dark* contrast corresponds to one orientation of magnetization and light contrast the opposite. *Reprinted with permission from Ref.* [87]. *Copyright 2014 by the Nature Publishing Group*

Subsequently, Lambert et al. demonstrated optical control of ferromagnetic materials ranging from magnetic thin films to multilayers and even granular films being explored for ultrahigh-density magnetic recording [88]. In a series of [Co (0.4 nm)/Pt(0.7)]$_N$ multilayers with N = 3, 5, and 8, the magneto-optical response to circularly polarized light of two helicities and linearly polarized light showed domain formation in the region scanned by the laser independent of light polarization representing TD for the thickest multilayer stack (N = 8). The situation remained qualitatively similar with an increase in average domain size for N = 5, but a fully deterministic magnetization reversal is observed for N = 5 for both circular polarizations, a clear demonstration of AO-HDS in a ferromagnetic material. It was further observed that although linear polarization leads to domain formation in zero applied field, the application of a magnetic field can stabilize a uniform magnetization state. This field increases from 3 to 4 Oe for N = 3 repeats, to ~12 Oe for N = 5 repeats, and to ~40 Oe for N = 8 repeats demonstrating the increased dipolar energy with thickness. When the applied field is combined with circular polarization, the applied field can either support or oppose the circular polarization.

The generality of this phenomenon is demonstrated by studying a range of ferromagnetic materials such as [Co(t_{Co})/Pt(t_{Pt})]$_N$, [Co(t_{Co})/Pd(t_{Pt})]$_N$, [Co$_x$Ni$_{1-x}$(0.6 nm)/Pt(0.7 nm)]$_N$, and [Co/Ni]$_N$ multilayer structures, where several material parameters (t_{Co}, t_{Pt}, N, and Ni concentration) were varied that change materials' magnetic properties, such as magnetization, T_C, anisotropy, and exchange interaction. The AO-HDS was further studied in combination with applied magnetic fields on high-anisotropy FePt-based HAMR media, which are FePtAgC and FePtC granular films that form high-anisotropy FePt grains separated by C grain boundaries. The average grain size was ~9.7 and ~7.7 nm for the FePtAgC and FePtC, respectively, with the room-temperature magnetic anisotropy and coercive fields of 7 and 3.5 T, respectively. Despite the lack of spatial resolution of the magneto-optical Faraday microscope, a net magnetization was achieved that depends on the helicity of the circularly polarized light, whereas no change is observed with linear polarization. That clearly demonstrated that the film magnetization was indeed controlled by the polarization of the light.

It is thus understood that AO-HDS is not exclusive to ferromagnetic materials, and therefore, antiferromagnetic exchange coupling between two magnetic sublattices is not required. Also, observation of AO-HDS switching on single Co films as well as Co/Pt and Co/Ni multilayers suggested that the requirements of superdiffusive currents that couple different magnetic regions in a heterogeneous sample are unlikely. The results, however, suggest that heating near the Curie point is important for the AO-HDS in ferromagnetic materials because the threshold intensities for both AO-HDS and TD generally track with the expected trends for T_C and do not scale with parameters such as interlayer exchange or anisotropy. Near T_C, the inverse Faraday effect or transfer of angular momentum from the light to the magnetic system is expected to be most effective. The experimental geometry causes the optomagnetic field to be parallel to the magnetization, which rules out the possibility of precessional switching well below T_C. Another important aspect of AO-HDS is that the magnetization state must be maintained during sample cooling after being switched near T_C. If demagnetization and thermal energies are too large, then the sample will demagnetize during cooling. For perpendicular magnetized films, there are strong dipolar fields within the film that supports domain formation. The dipolar energy gain for domain formation is strongly suppressed in the ultrathin film limit and explains the observation of AO-HDS only in the thin-film limit.

The coherent control of spin dynamics offers new and technologically important class of materials and opens new directions in integrated magnetic and optical memory, data storage, and processing applications. Further, progress toward a better understanding of the interaction between pulsed polarized light and magnetic materials is envisaged for the realization of real applications.

References

1. Jánossy A, Monod P (1976) Investigation of magnetic coupling at the interface of a ferromagnetic and paramagnetic metal by conduction electron spin resonance. Solid State Commun 18(2):203–205. doi:http://dx.doi.org/10.1016/0038-1098(76)91453-8
2. Silsbee RH, Janossy A, Monod P (1979) Coupling between ferromagnetic and conduction-spin-resonance modes at a ferromagnetic-normal-metal interface. Phys Rev B 19(9):4382–4399. doi:10.1103/PhysRevB.19.4382
3. Grünberg P, Schreiber R, Pang Y, Brodsky MB, Sowers H (1986) Layered magnetic structures: evidence for antiferromagnetic coupling of Fe layers across Cr interlayers. Phys Rev Lett 57(19):2442–2445. doi:10.1103/PhysRevLett.57.2442
4. Baibich MN, Broto JM, Fert A, Vandau FN, Petroff F, Eitenne P, Creuzet G, Friederich A, Chazelas J (1988) Giant magnetoresistance of (001)Fe/(001) Cr magnetic superlattices. Phys Rev Lett 61(21):2472–2475. doi:10.1103/PhysRevLett.61.2472
5. Binasch G, Grünberg P, Saurenbach F, Zinn W (1989) Enhanced magnetoresistance in layered magnetic-structures with antiferromagnetic interlayer exchange. Phys Rev B 39(7): 4828–4830. doi:10.1103/PhysRevB.39.4828
6. Parkin SSP, More N, Roche KP (1990) Oscillations in exchange coupling and magnetoresistance in metallic superlattice structures—Co/Ru, Co/Cr, and Fe/Cr. Phys Rev Lett 64 (19):2304–2307. doi:10.1103/PhysRevLett.64.2304
7. Parkin SSP, Bhadra R, Roche KP (1991) Oscillatory magnetic exchange coupling through thin copper layers. Phys Rev Lett 66(16):2152–2155. doi:10.1103/PhysRevLett.66.2152
8. Parkin SSP, Kaiser C, Panchula A, Rice PM, Hughes B, Samant M, Yang S-H (2004) Giant tunnelling magnetoresistance at room temperature with MgO (100) tunnel barriers. Nat Mater 3(12):862–867. doi:10.1038/nmat1256
9. Yuasa S, Nagahama T, Fukushima A, Suzuki Y, Ando K (2004) Giant room-temperature magnetoresistance in single-crystal Fe/MgO/Fe magnetic tunnel junctions. Nat Mater 3(12): 868–871. doi:10.1038/nmat1257
10. Dieny B, Speriosu VS, Gurney BA, Parkin SSP, Wilhoit DR, Roche KP, Metin S, Peterson DT, Nadimi S (1991) Spin-valve effect in soft ferromagnetic sandwiches. J Magn Magn Mater 93:101–104. doi:10.1016/0304-8853(91)90311-w
11. Suezawa Y, Gondo Y (1993) Spin-polarized electrons and magnetoresistance in ferromagnetic tunnel-junctions and multilayers. J Magn Magn Mater 126(1–3):524–526. doi:10.1016/0304-8853(93)90677-t

12. Parkin SSP (1993) Origin of enhanced magnetoresistance of magnetic multilayers—spin-dependent scattering from magnetic interface states. Phys Rev Lett 71(10):1641–1644. doi:10.1103/PhysRevLett.71.1641
13. Kirilyuk A, Kimel AV, Rasing T (2010) Ultrafast optical manipulation of magnetic order. Rev Mod Phys 82(3):2731–2784. doi:https://doi.org/10.1103/RevModPhys.82.2731
14. Berger L (1996) Emission of spin waves by a magnetic multilayer traversed by a current. Phys Rev B 54(13):9353–9358. doi:10.1103/PhysRevB.54.9353
15. Slonczewski JC (1996) Current-driven excitation of magnetic multilayers. J Magn Magn Mater 159(1–2):L1–L7. doi:10.1016/0304-8853(96)00062-5
16. Tsoi M, Jansen AGM, Bass J, Chiang WC, Seck M, Tsoi V, Wyder P (1998) Excitation of a magnetic multilayer by an electric current. Phys Rev Lett 80(19):4281–4284. doi:10.1103/PhysRevLett.80.4281
17. Myers EB, Ralph DC, Katine JA, Louie RN, Buhrman RA (1999) Current-induced switching of domains in magnetic multilayer devices. Science 285(5429):867–870. doi:10.1126/science.285.5429.867
18. Thomas L, Hayashi M, Jiang X, Moriya R, Rettner C, Parkin SSP (2006) Oscillatory dependence of current-driven magnetic domain wall motion on current pulse length. Nature 443(7108):197–200. doi:10.1038/nature05093
19. Parkin SSP, Hayashi M, Thomas L (2008) Magnetic domain-wall racetrack memory. Science 320(5873):190–194. doi:10.1126/science.1145799
20. Hayashi M, Thomas L, Moriya R, Rettner C, Parkin SSP (2008) Current-controlled magnetic domain-wall nanowire shift register. Science 320(5873):209–211. doi:10.1126/science.1154587
21. Mihai Miron I, Gaudin G, Auffret S, Rodmacq B, Schuhl A, Pizzini S, Vogel J, Gambardella P (2010) Current-driven spin torque induced by the Rashba effect in a ferromagnetic metal layer. Nat Mater 9(3):230–234. doi:10.1038/nmat2613
22. Mihai Miron I, Moore T, Szambolics H, Buda-Prejbeanu LD, Auffret S, Rodmacq B, Pizzini S, Vogel J, Bonfim M, Schuhl A, Gaudin G (2011) Fast current-induced domain-wall motion controlled by the Rashba effect. Nat Mater 10(6):419–423. doi:10.1038/nmat3020
23. Liu L, Pai C-F, Li Y, Tseng HW, Ralph DC, Buhrman RA (2012) Spin-torque switching with the giant spin Hall effect of Tantalum. Science 336(6081):555–558. doi:10.1126/science.1218197
24. Haazen PPJ, Mure E, Franken JH, Lavrijsen R, Swagten HJM, Koopmans B (2013) Domain wall depinning governed by the spin Hall effect. Nat Mater 12(4):299–303. doi:10.1038/nmat3553
25. Torrejon J, Kim J, Sinha J, Mitani S, Hayashi M, Yamanouchi M, Ohno H (2014) Interface control of the magnetic chirality in CoFeB/MgO heterostructures with heavy-metal underlayers. Nat Commun 5:4655. doi:10.1038/ncomms5655
26. Ralph DC, Stiles MD (2008) Spin transfer torques. J Magn Magn Mater 320(7):1190–1216. doi:http://dx.doi.org/10.1016/j.jmmm.2007.12.019
27. Slonczewski JC (1999) Excitation of spin waves by an electric current. J Magn Magn Mater 195(2):L261–L268. doi:10.1016/s0304-8853(99)00043-8
28. Katine JA, Albert FJ, Buhrman RA, Myers EB, Ralph DC (2000) Current-driven magnetization reversal and spin-wave excitations in Co/Cu/Co pillars. Phys Rev Lett 84(14):3149–3152. doi:10.1103/PhysRevLett.84.3149
29. Kiselev SI, Sankey JC, Krivorotov IN, Emley NC, Schoelkopf RJ, Buhrman RA, Ralph DC (2003) Microwave oscillations of a nanomagnet driven by a spin-polarized current. Nature 425(6956):380–383. doi:10.1038/nature01967
30. Piraux L, Renard K, Guillemet R, Mátéfi-Tempfli S, Mátéfi-Tempfli M, Antohe VA, Fusil S, Bouzehouane K, Cros V (2007) Template grown NiFe/Cu/NiFe nanowires for spin transfer devices. Nano Lett 7(9):2563–2567. doi:10.1021/nl070263s

31. Matsumoto R, Fukushima A, Yakushiji K, Yakata S, Nagahama T, Kubota H, Katayama T, Suzuki Y, Ando K, Yuasa S, Georges B, Cros V, Grollier J, Fert A (2009) Spin torque-induced switching and precession in fully epitaxial Fe/MgO/Fe magnetic tunnel junctions. Phys Rev B 80(17):174405. doi:https://doi.org/10.1103/PhysRevB.80.174405

32. Krivorotov IN, Berkov DV, Gorn NL, Emley NC, Sankey JC, Ralph DC, Buhrman RA (2007) Large-amplitude coherent spin waves excited by spin-polarized current in nanoscale spin valves. Phys Rev B 76(2):024418. doi:10.1103/PhysRevB.76.024418

33. Houssameddine D, Ebels U, Delaet B, Rodmacq B, Firastrau I, Ponthenier F, Brunet M, Thirion C, Michel JP, Prejbeanu-Buda L, Cyrille MC, Redon O, Dieny B (2007) Spin-torque oscillator using a perpendicular polarizer and a planar free layer. Nat Mater 6(6):447–453. doi:10.1038/nmat1905

34. Thadani KV, Finocchio G, Li ZP, Ozatay O, Sankey JC, Krivorotov IN, Cui YT, Buhrman RA, Ralph DC (2008) Strong linewidth variation for spin-torque nano-oscillators as a function of in-plane magnetic field angle. Phy Rev B 78(2):024409. doi:https://doi.org/10.1103/PhysRevB.78.024409

35. Bosu S, Sepehri-Amin H, Sakuraba Y, Hayashi M, Abert C, Suess D, Schrefl T, Hono K (2016) Reduction of critical current density for out-of-plane mode oscillation in a mag-flip spin torque oscillator using highly spin-polarized $Co_2Fe(Ga_{0.5}Ge_{0.5})$ spin injection layer. Appl Phys Lett 108(7):072403. doi:10.1063/1.4942373

36. Sinha J, Hayashi M, Takahashi YK, Taniguchi T, Drapeko M, Mitani S, Hono K (2011) Large amplitude microwave emission and reduced nonlinear phase noise in Co_2Fe $(Ge_{0.5}Ga_{0.5})$ Heusler alloy based pseudo spin valve nanopillars. Appl Phys Lett 99 (16):162508. doi:10.1063/1.3647771

37. Pribiag VS, Krivorotov IN, Fuchs GD, Braganca PM, Ozatay O, Sankey JC, Ralph DC, Buhrman RA (2007) Magnetic vortex oscillator driven by D.C. spin-polarized current. Nat Phys 3(7):498–503. doi:10.1038/nphys619

38. Slavin A, Tiberkevich V (2005) Spin wave mode excited by spin-polarized current in a magnetic nanocontact is a standing self-localized wave bullet. Phys Rev Lett 95(23):237201. doi:10.1103/PhysRevLett.95.237201

39. Rippard WH, Pufall MR, Kaka S, Silva TJ, Russek SE (2004) Current-driven microwave dynamics in magnetic point contacts as a function of applied field angle. Phys Rev B 70 (10):100406. doi:10.1103/PhysRevB.70.100406

40. Rippard WH, Pufall MR, Kaka S, Silva TJ, Russek SE, Katine JA (2005) Injection locking and phase control of spin transfer nano-oscillators. Phys Rev Lett 95(6):067203. doi:10.1103/PhysRevLett.95.067203

41. Kim J-V, Tiberkevich V, Slavin AN (2008) Generation linewidth of an auto-oscillator with a nonlinear frequency shift: spin-torque nano-oscillator. Phys Rev Lett 100(1):017207. doi:10.1103/PhysRevLett.100.017207

42. Okura R, Sakuraba Y, Seki T, Izumi K, Mizuguchi M, Takanashi K (2011) High-power Rf oscillation induced in half-metallic Co_2MnSi layer by spin-transfer torque. Appl Phys Lett 99 (5):052510. doi:10.1063/1.3624470

43. Garello K, Miron IM, Avci CO, Freimuth F, Mokrousov Y, Bluegel S, Auffret S, Boulle O, Gaudin G, Gambardella P (2013) Symmetry and magnitude of spin-orbit torques in ferromagnetic heterostructures. Nat Nanotechnol 8(8):587–593. doi:10.1038/nnano.2013.145

44. Johnson M, Silsbee RH (1985) Interfacial charge-spin coupling: injection and detection of spin magnetization in metals. Phys Rev Lett 55(17):1790–1793. doi:10.1103/PhysRevLett.55.1790

45. Dyakonov MI, Perel VI (1971) Current-induced spin orientation of electrons in semiconductors. Phys Lett A 35(6):459–460. doi:http://dx.doi.org/10.1016/0375-9601(71)90196-4

46. Manchon A, Zhang S (2009) Theory of spin torque due to spin-orbit coupling. Phys Rev B 79 (9):094422. doi:10.1103/PhysRevB.79.094422

47. Yu AB, Rashba EI (1984) Oscillatory effects and the magnetic susceptibility of carriers in inversion layers. J Phys C: Solid State Phys 17(33):6039. doi:10.1088/0022-3719/17/33/015
48. Liu L, Lee OJ, Gudmundsen TJ, Ralph DC, Buhrman RA (2012) Current-induced switching of perpendicularly magnetized magnetic layers using spin torque from the spin Hall effect. Phys Rev Lett 109(9). doi:10.1103/PhysRevLett.109.096602
49. Kim J, Sinha J, Hayashi M, Yamanouchi M, Fukami S, Suzuki T, Mitani S, Ohno H (2013) Layer thickness dependence of the current-induced effective field vector in Ta|CoFeB|MgO. Nat Mater 12(3):240–245. doi:10.1038/nmat3522
50. Edelstein VM (1990) Spin polarization of conduction electrons induced by electric current in two-dimensional asymmetric electron systems. Solid State Commun 73(3):233–235. doi:10.1016/0038-1098(90)90963-C
51. Hirsch JE (1999) Spin Hall effect. Phys Rev Lett 83(9):1834–1837. doi:10.1103/PhysRevLett.83.1834
52. Manchon A, Zhang S (2008) Theory of nonequilibrium intrinsic spin torque in a single nanomagnet. Phys Rev B 78(21):212405. doi:10.1103/PhysRevB.78.212405
53. Hoffmann A (2013) Spin Hall effects in metals. IEEE Trans Magn 49(10):5172–5193. doi:10.1109/tmag.2013.2262947
54. Sinova J, Valenzuela SO, Wunderlich J, Back CH, Jungwirth T (2015) Spin Hall effects. Rev Mod Phys 87(4):1213–1260. doi:10.1103/RevModPhys.87.1213
55. Saitoh E, Ueda M, Miyajima H, Tatara G (2006) Conversion of spin current into charge current at room temperature: inverse spin-Hall effect. Appl Phys Lett 88(18):182509. doi:10.1063/1.2199473
56. Kimura T, Otani Y, Sato T, Takahashi S, Maekawa S (2007) Room-temperature reversible spin Hall effect. Phys Rev Lett 98(15):156601. doi:10.1103/PhysRevLett.98.156601
57. Tanaka T, Kontani H, Naito M, Naito T, Hirashima DS, Yamada K, Inoue J (2008) Intrinsic spin Hall effect and orbital Hall effect in $4d$ And $5d$ transition metals. Phys Rev B 77 (16):165117. doi:10.1103/PhysRevB.77.165117
58. Morota M, Niimi Y, Ohnishi K, Wei DH, Tanaka T, Kontani H, Kimura T, Otani Y (2011) Indication of intrinsic spin Hall effect in 4d and 5d transition metals. Phys Rev B 83(17). doi:10.1103/PhysRevB.83.174405
59. Haney PM, Lee H-W, Lee K-J, Manchon A, Stiles MD (2013) Current induced torques and interfacial spin-orbit coupling: semiclassical modeling. Phys Rev B 87(17):174411. doi:10.1103/PhysRevB.87.174411
60. Haney PM, Lee H-W, Lee K-J, Manchon A, Stiles MD (2013) Current-induced torques and interfacial spin-orbit coupling. Phys Rev B 88(21):214417. doi:10.1103/PhysRevB.88.214417
61. Freimuth F, Blügel S, Mokrousov Y (2014) Spin-orbit torques in Co/Pt(111) And Mn/W(001) magnetic bilayers from first principles. Phys Rev B 90(17):174423. doi:10.1103/PhysRevB.90.174423
62. Zhang S, Levy PM, Fert A (2002) Mechanisms of spin-polarized current-driven magnetization switching. Phys Rev Lett 88(23):236601. doi:10.1103/PhysRevLett.88.236601
63. Shpiro A, Levy PM, Zhang S (2003) Self-consistent treatment of nonequilibrium spin torques in magnetic multilayers. Phys Rev B 67(10):104430. doi:10.1103/PhysRevB.67.104430
64. Stiles MD, Zangwill A (2002) Anatomy of spin-transfer torque. Phys Rev B 66(1):014407. doi:10.1103/PhysRevB.66.014407
65. Freeman MR, Brady MJ, Smyth J (1992) Extremely high frequency pulse magnetic resonance by picosecond magneto-optic sampling. Appl Phys Lett 60(20):2555–2557. doi:10.1063/1.106911
66. Freeman MR (1994) Picosecond pulsed-field probes of magnetic systems. J Appl Phys 75 (10):6194–6198. doi:10.1063/1.355454
67. Elezzabi AY, Freeman MR, Johnson M (1996) Direct measurement of the conduction electron spin-lattice relaxation time T_1 in gold. Phys Rev Lett 77(15):3220–3223. doi:10.1103/PhysRevLett.77.3220

68. Hiebert WK, Stankiewicz A, Freeman MR (1997) Direct observation of magnetic relaxation in a small permalloy disk by time-resolved scanning Kerr microscopy. Phys Rev Lett 79 (6):1134–1137. doi:10.1103/PhysRevLett.79.1134

69. Stotz JAH, Freeman MR (1997) A stroboscopic scanning solid immersion lens microscope. Rev Sci Instrum 68(12):4468–4477. doi:10.1063/1.1148416

70. Bauer M, Lopusnik R, Fassbender J, Hillebrands B, Dotsch H (2000) Successful suppression of magnetization precession after short field pulses. IEEE Trans Magn 36(5):2764–2766. doi:10.1109/20.908583

71. Bauer M, Lopusnik R, Fassbender J, Hillebrands B (2000) Suppression of magnetic-field pulse-induced magnetization precession by pulse tailoring. Appl Phys Lett 76(19):2758–2760. doi:10.1063/1.126466

72. Bauer M, Lopusnik R, Dötsch H, Kalinikos BA, Patton CE, Fassbender J, Hillebrands B (2001) Time domain MOKE detection of spin-wave modes and precession control for magnetization switching in ferrite films. J Magn Magn Mater 226–230, Part 1:507–509. doi:10.1016/S0304-8853(00)00992-6

73. Crawford TM, Kabos P, Silva TJ (2000) Coherent control of precessional dynamics in thin film permalloy. Appl Phys Lett 76(15):2113–2115. doi:10.1063/1.126280

74. Schumacher HW, Chappert C, Crozat P, Sousa RC, Freitas PP, Bauer M (2002) Coherent suppression of magnetic ringing in microscopic spin valve elements. Appl Phys Lett 80 (20):3781–3783. doi:10.1063/1.1480476

75. Schumacher HW, Chappert C, Crozat P, Sousa RC, Freitas PP, Miltat J, Fassbender J, Hillebrands B (2003) Phase coherent precessional magnetization reversal in microscopic spin valve elements. Phys Rev Lett 90(1):017201. doi:10.1103/PhysRevLett.90.017201

76. Gerrits T, van den Berg HAM, Hohlfeld J, Bar L, Rasing T (2002) Ultrafast precessional magnetization reversal by picosecond magnetic field pulse shaping. Nature 418(6897):509–512. doi:10.1038/nature00905

77. Barman A, Kruglyak VV, Hicken RJ, Scott J, Rahman M (2005) Dependence of spatial coherence of coherent suppression of magnetization precession upon aspect ratio in $Ni_{81}Fe_{19}$ microdots. J Appl Phys 97(10):10A710. doi:10.1063/1.1850834

78. Kimel AV, Kirilyuk A, Usachev PA, Pisarev RV, Balbashov AM, Rasing T (2005) Ultrafast non-thermal control of magnetization by instantaneous photomagnetic pulses. Nature 435 (7042):655–657. doi:10.1038/nature03564

79. Kirilyuk A, Kimel AV, Rasing T (2011) Controlling spins with light. Philos Trans R Soc A: Math Phys Eng Sci 369(1951):3631–3645. doi:10.1098/rsta.2011.0168

80. Stanciu CD, Hansteen F, Kimel AV, Kirilyuk A, Tsukamoto A, Itoh A, Rasing T (2007) All-optical magnetic recording with circularly polarized light. Phys Rev Lett 99(4). doi:10.1103/PhysRevLett.99.047601

81. Radu I, Vahaplar K, Stamm C, Kachel T, Pontius N, Durr HA, Ostler TA, Barker J, Evans RFL, Chantrell RW, Tsukamoto A, Itoh A, Kirilyuk A, Rasing T, Kimel AV (2011) Transient ferromagnetic-like state mediating ultrafast reversal of antiferromagnetically coupled spins. Nature 472(7342):205–208. doi:10.1038/nature09901

82. Graves CE, Reid AH, Wang T, Wu B, de Jong S, Vahaplar K, Radu I, Bernstein DP, Messerschmidt M, Mueller L, Coffee R, Bionta M, Epp SW, Hartmann R, Kimmel N, Hauser G, Hartmann A, Holl P, Gorke H, Mentink JH, Tsukamoto A, Fognini A, Turner JJ, Schlotter WF, Rolles D, Soltau H, Strueder L, Acremann Y, Kimel AV, Kirilyuk A, Rasing T, Stoehr J, Scherz AO, Duerr HA (2013) Nanoscale spin reversal by non-local angular momentum transfer following ultrafast laser excitation in ferrimagnetic GdFeCo. Nat Mater 12(4):293–298. doi:10.1038/nmat3597

83. Turgut E, Grychtol P, La-O-Vorakiat C, Adams DE, Kapteyn HC, Murnane MM, Mathias S, Aeschlimann M, Schneider CM, Shaw JM, Nembach HT, Silva TJ (2013) Publisher's note: reply to "Comment on 'Ultrafast demagnetization measurements using extreme ultraviolet light: comparison of electronic and magnetic contributions'" [Phys. Rev. X 3, 038002 (2013) PRXHAE2160-3308]. Phys Rev X 3(3), 039901. doi:10.1103/PhysRevX.3.039901

84. Turgut E, La-o-vorakiat C, Shaw JM, Grychtol P, Nembach HT, Rudolf D, Adam R, Aeschlimann M, Schneider CM, Silva TJ, Murnane MM, Kapteyn HC, Mathias S (2013) Controlling the competition between optically induced ultrafast spin-flip scattering and spin transport in magnetic multilayers. Phys Rev Lett 110(19):197201. doi:10.1103/PhysRevLett. 110.197201
85. Battiato M, Carva K, Oppeneer PM (2010) Superdiffusive spin transport as a mechanism of ultrafast demagnetization. Phys Rev Lett 105(2):027203. doi:10.1103/PhysRevLett.105. 027203
86. Malinowski G, Longa FD, Rietjens JHH, Paluskar PV, Huijink R, Swagten HJM, Koopmans B (2008) Control of speed and efficiency of ultrafast demagnetization by direct transfer of spin angular momentum. Nat Phys 4(11):855–858. doi:10.1038/nphys1092
87. Mangin S, Gottwald M, Lambert CH, Steil D, Uhlir V, Pang L, Hehn M, Alebrand S, Cinchetti M, Malinowski G, Fainman Y, Aeschlimann M, Fullerton EE (2014) Engineered materials for all-optical helicity-dependent magnetic switching. Nat Mater 13(3):287–293. doi:10.1038/nmat3864
88. Lambert C.-H, Mangin S, Varaprasad BSDCS, Takahashi YK, Hehn M, Cinchetti M, Malinowski G, Hono K, Fainman Y, Aeschlimann M, Fullerton EE (2014) All-optical control of ferromagnetic thin films and nanostructures. Science 345(6202):1337–1340. doi:10.1126/ science.1253493

Chapter 7
Tunable Magnetic Damping in Ferromagnetic/Non-magnetic Bilayer Films

With the advancement in the methods for thin film deposition, tailoring of magnetic properties of thin ferromagnetic layer by using a neighboring layer has gained attention of researchers. From the perspective of fundamental understanding, the role of interface in different bilayer and multilayer systems has been an intense area of investigation. Utilizing the additional tunable properties originating from the interface has been foreseen as a route to improve the performance of numerous devices in application. In this chapter, the important aspect of magnetic damping in ferromagnetic/non-magnetic (FM/NM) bilayer films with primary focus on the recent experimental results is presented. Before describing various cases, we discuss two important mechanisms involved in FM/NM bilayer system, namely spin pumping and interfacial *d-d* hybridization.

7.1 Static Control of Damping

Controlling magnetic damping in FM/NM bilayer system is an important requirement for developing advanced spintronics-based devices, and it is one of the most active areas of modern spintronics research. Damping plays a crucial role in spin transfer torque magnetoresistive random access memory (STT-MRAM) devices [1] and also in magnonic devices [2]. As briefly mentioned in Chap. 1, the low damping is known to facilitate a lower write current in STT-MRAM devices and longer propagation of spin waves in magnonic devices, whereas higher damping is desirable for increasing the magnetization reversal rates and the coherent reversal of magnetic elements. In Chap. 3, we have already described that the total damping is the sum of intrinsic and extrinsic contributions.

© Springer International Publishing AG 2018
A. Barman and J. Sinha, *Spin Dynamics and Damping in Ferromagnetic Thin Films and Nanostructures*, https://doi.org/10.1007/978-3-319-66296-1_7

7.1.1 Spin Pumping

Spin pumping was originally proposed by Berger in 1996 [3], and the theory was further elaborated by Brataas et al. in 2002 [4]. Considering FM/NM bilayers, in these theoretical studies, it was predicted that the presence of the ferromagnet induces a small, decaying, oscillatory magnetization in the neighboring non-magnetic layer. The origin of this magnetization was considered to be arising from the partial reflection of electrons while scattering from the interface. The incoming and outgoing components of the electron states interfere with each other and ultimately give rise to net induced oscillatory spin density which subsequently decays in the non-magnet. Overall, this phenomenon can be viewed as the concept for spin current injection into arbitrary conductors through Ohmic contacts, which does not involve net charge currents. The spin source is a ferromagnetic reservoir at resonance with an rf field. The precessing magnetization gives rise to spin current transported into an adjacent normal metal at the cost of its own angular momentum. In FM/NM bilayers, the magnetic damping increases if the spin current is allowed to leak through an adjacent non-magnetic material. In other words, an external excitation induces magnetization precession due to which angular momentum is transferred from ferromagnetic to the non-magnetic layer in the form of spin current which is analogous to a physical pumping mechanism, hence known as spin pumping effect. The magnetization precession can be described by the Landau–Lifshitz–Gilbert (LLG) equation described in previous chapters. The spin pumping term has exactly the same form as the Gilbert damping [5]. The spin absorption efficiency can be estimated by the inverse of spin-flip relaxation time (τ_{SF}) which is proportional to Z^4, where Z is the atomic number. Hence, a stronger spin pumping is expected in the presence of heavy metal layer adjacent to the ferromagnet. Tserkovnyak et al., in 2002 [6], theoretically calculated spin pumping and explained the behavior of damping using time-dependent scattering theory in case of ferromagnetic sandwich structures (NM/FM/NM) as experimentally obtained from FMR line width by Mizukami et al. in 2001 [7–9]. The spin current can be expressed in the form of the equation below:

$$I_S^{\text{Pump}} = \frac{\hbar}{4\pi}\left(A_r m \times \frac{dm}{dt} - A_i \frac{dm}{dt}\right) \tag{7.1}$$

where scattering parameters A_r and A_i are expressed as

$$A_r = 1/2 \sum_{mn}\left\{\left|r_{mn}^{\uparrow} - r_{mn}^{\downarrow}\right|^2 + \left|t_{mn}^{\uparrow} - t_{mn}^{\downarrow}\right|^2\right\} \tag{7.2}$$

and

$$A_i = Im \sum_{mn} \left\{ r^{\uparrow}_{mn} \left(r^{\downarrow}_{mn} \right)^* + t^{\uparrow}_{mn} \left(t^{\downarrow}_{mn} \right)^* \right\} \tag{7.3}$$

here, $r^{\uparrow}_{mn}$ and $r^{\downarrow}_{mn}$ are reflection coefficients, whereas $t^{\uparrow}_{mn}$ and $t^{\downarrow}_{mn}$ are transmission coefficients for up and down electrons. Considering the conservation of angular momentum in case of spin pumping, the magnetization dynamics can be represented by the LLG equation where damping parameter (α) and gyromagnetic ratio (γ) are modified as:

$$\alpha = \frac{\gamma}{\gamma_0} \left[\alpha_0 + g \left(A^C_i + A^S_i \right) / 4\pi M \right] \tag{7.4}$$

and

$$\frac{1}{\gamma} = \frac{1}{\gamma_0} \left[1 + g \left(A^C_i + A^S_i \right) / 4\pi M \right] \tag{7.5}$$

where the superscripts 'C' and 'S' indicate the capping and seed normal metal layer, and the subscript '0' denotes the bulk value.

In order to get a qualitative picture of the spin pumping, let us consider density of states of conduction electrons in a ferromagnet during spin pumping (see Fig. 7.1). Initially, the bands corresponding to spin up and spin down electrons (red and blue) are filled up to Fermi level with relative shift equal to exchange energy (Fig. 7.1 left). A change in magnetization orientation causes a further shift in the energy level by allowing electrons above Fermi energy to relax via spin-flip process to fill the lower energy spin band. An additional relaxation process takes place in the presence of an adjacent non-magnetic layer via spin pumping followed by a spin flip in the non-magnetic layer (Fig. 7.1 middle) until the equilibrium is achieved (Fig. 7.1 right).

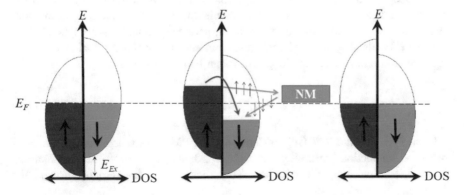

Fig. 7.1 Schematic illustration of spin pumping using density of states

Brataas et al. in 2002 proposed circuit theory which is based on the approximation that FM/NM bilayer can be divided into nodes and regions that connect them [4]. In the nodes, the voltages (chemical potentials) are assumed to be spatially uniform spin accumulation. Furthermore, it is presumed that the nodes are connected by conducting channels, across which the chemical potential differences drive charge and spin currents. By invoking the concept of spin-mixing conductance (mathematical form will be discussed in later section) that describes the behavior of spins in the non-magnet that are perpendicular to the magnetization in the ferromagnet, they developed the theory of spin pumping. Through their study, it was concluded that the real part of the spin-mixing conductance gives the spin current that is aligned with the perpendicular part of the chemical potential, whereas the imaginary part gives the spin current that is perpendicular to both the magnetization and the chemical potential in the non-magnet. Interestingly, both of these components are absorbed at the interface giving rise to a torque on the magnetization. Furthermore, due to spin filtering and dephasing, the sum over the reflection amplitudes vanishes. Particularly, in case of transition metals, the real part of the mixing conductance is roughly proportional to the number of conducting channels, and the imaginary part is close to zero.

7.1.2 Interfacial d-d Hybridization

Among various microscopic theories of Gilbert damping, Kambersky's model developed in 1976 suggests that the Gilbert damping originates from spin–orbit scattering of band electrons in ferromagnetic metals [10]. A crude approximation for damping within this model yields following expression

$$\alpha \approx \frac{1}{\gamma M_S} \left[\frac{\mu_B^2 D(E_F)}{\tau_E} \right] \left(\frac{\xi}{W} \right)^2 \tag{7.6}$$

where $D(E_F)$ is the density of states at the Fermi level for d-bands, M_s is the saturation magnetization, μ_B is Bohr magneton, τ_e is the electron momentum relaxation time, ξ is the spin–orbit coupling parameter, and W is the typical band width of d-bands. The perpendicular magnetic anisotropy (PMA) is also known to originate from spin–orbit coupling, and it can be expressed as

$$K_U \approx \xi \left(\frac{\xi}{W} \right) \tag{7.7}$$

From Eqs. 7.6 and 7.7, Gilbert damping and PMA are second-order effects of spin–orbit interaction perturbing the d-electron band structure. In 1998, Nakajima et al. [11] using magnetic circular X-ray dichroism measurement showed the strongly enhanced perpendicular Co orbital moment (m_{orb}) in Co/Pt multilayers exhibiting PMA which originates from Co $3d$-Pt $5d$ interfacial hybridization.

Mizukami et al. in 2010 [13] investigated the magnetization dynamics in perpendicularly magnetized Pt/Co/Pt layers and discussed the possibility of Co $3d$–Pt $5d$ hybridization effect, which decreases the bandwidth W for the Co atomic layer in contact with a Pt layer, enhancing both PMA and Gilbert damping. In 2011, Pal et al. [12] investigated magnetization dynamics in [Co(t_{Co})/Pd(0.9 nm)$_8$] multilayers with t_{Co} varying between 1 and 0.22 nm. Using vibrating sample magnetometer (VSM) and static polar magneto-optical Kerr effect measurements, all samples were found to show PMA. The anisotropy field increases systematically with decrease in t_{Co} and exhibits a maximum at 0.22 nm, beyond which it decreases sharply. The saturation magnetization decreases monotonically with the decrease in t_{Co} over the entire range down to 0.13 nm. Using time-resolved magneto-optical Kerr effect study, the damping coefficient α was extracted. Interestingly, α was found to be linearly proportional to K_{eff} with a slope of 4.33×10^{-8} cc/erg as reproduced in Fig. 7.2. They considered various possible mechanisms for the enhancement of α and concluded that magnon–magnon scattering is less effective as these samples are perpendicularly magnetized. The spin pumping effect is also found to be negligible by carefully analyzing the slope of the relaxation frequency versus inverse of Co thickness plot. The possibility of interface roughness and alloying was also ruled out as it may cause the increase in α, while the K_{eff} will decrease. Subsequently, they explained the increase in α with K_{eff} due to interfacial d-d hybridization at the Co-Pt interface. Though the mechanism of d-d hybridization has been invoked in several experimental studies in order to explain the enhanced value of damping, the detailed theoretical study for d-d hybridization in the context of its implication on Gilbert damping is limited to few studies. Most theoretical and experimental studies regarding d-d hybridization have been related to the explanation of PMA in the FM/NM systems. There have been further attempts to provide a generic overview on the correlation of PMA with the damping; however, up to date, there has not been a clear understanding due to variation in results obtained for different samples and interpretation provided by various research groups.

Fig. 7.2 The damping coefficient α is plotted as a function of K_{eff} (symbols) and the dotted line corresponds to the linear fit. *Reprinted with permission from Ref. [12]. Copyright 2011 by American Institute of Physics*

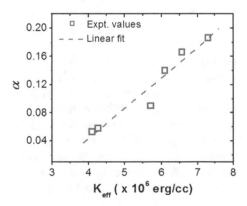

7.1.3 Exemplary Description of Damping Studies in Co and NiFe Layer with Pt or Au Overlayer

During last few years, the influence of non-magnetic heavy metal (NM) seed layer or over layer on the damping in adjacent ferromagnetic (FM) layers has attracted a lot of research interest [9, 14–17]. In such bilayer systems, there are several mechanisms that may lead to the enhancement of damping. Briefly, spin–orbit coupling (SOC) and interfacial d-d hybridization enhance the intrinsic damping, while extrinsic enhancement of the damping can arise from two-magnon scattering processes, linked to roughness and defects at the interface region. As described in Chap. 3, the total precessional damping is a sum of the intrinsic and extrinsic contributions.

In 2016, Azzawi et al. [18] studied the magnetization dynamics using time-resolved magneto-optical Kerr effect (TR-MOKE, an all-optical pump-probe technique) in Co and $Ni_{81}Fe_{19}$ with NM capping layers of Pt and Au. Bilayer thin films were sputter-deposited on to thermally oxidized silicon substrate with a 100 nm SiO_2 layer. The Co layer thicknesses in their study were either 4 or 10 nm, and for the $Ni_{81}Fe_{19}$ films, the thickness was 7 nm whereas the capping layer thicknesses were varied from 0.2 to 10 nm. Through structural study, it was found that both Pt and Au form a continuous layer on Co at a thickness greater than ~ 0.7 nm, while for $Ni_{81}Fe_{19}$ the NM capping layer becomes continuous at thicknesses above ~ 0.9 nm. The growth of Au or Pt capping layers onto a FM thin film layer begins with the formation of localized islands of the NM material; these expand and develop into a continuous capping layer as the NM thickness increases. For Au and Pt capping of Co, the interfacial roughness is very similar; however, details of the local atomic arrangement differ, and Co and Pt are miscible, while Co and Au are immiscible. In this study, they aimed to examine the effect of sub-nanometer thickness of specific NM materials on the damping.

In Fig. 7.3a and c, the variation of effective damping α_{eff} with non-magnetic layer (Pt and Au) thickness and, in Fig. 7.3b, precession frequency and saturation magnetization with non-magnetic layer thickness are reproduced from Azzawi et al. [18]. It may be noted from these figures that α_{eff} increases significantly for both Co/Pt and $Ni_{81}Fe_{19}$/Pt layers as the Pt capping layer thickness increases and peaks around 0.7–0.8 nm for Co/Pt and 0.6 nm for $Ni_{81}Fe_{19}$/Pt. However, in the case of Co/Au, α_{eff} is nearly constant across the entire Au thickness range. It is important to note here that in the discussed experiment, enhanced value of damping in comparison with that of hcp Co (0.011) is likely related to partially oxidized FM layers. The thickness dependence of the NM layer on α_{eff} for Co/Pt and $Ni_{81}Fe_{19}$/Pt may be divided into three different regions. With initial increasing Pt thickness, the damping first increases rapidly (region I) and then peaks (region II) before decreasing to a smaller value (region III). This complex dependence with three characteristic regions can be understood by considering several intrinsic and extrinsic effects occurring at the interface of the FM and NM layers. Also, the experimental result of Azzawi et al. [18] is in agreement with a recent theoretical

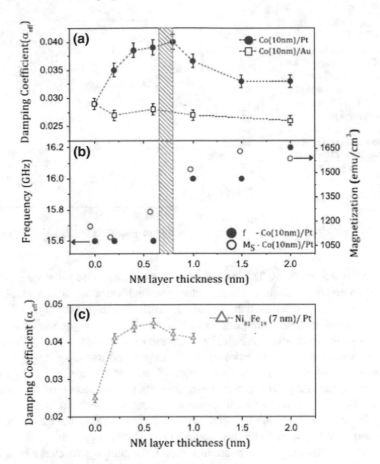

Fig. 7.3 **a** and **c** α_{eff} as a function of t_{NM} for Co/Pt and $Ni_{81}Fe_{19}$/Pt, respectively. **b** Frequency and saturation magnetization as a function of t_{NM} and ~ 1.4 kOe of magnetic field; it shows a similar trend in their variants. The shaded bar indicates the Pt thickness where the Pt became continuous. *Reprinted with permission from Ref.* [18]. *Copyright 2016 by American Physical Society*

study by Barati et al. [19] where damping increases with increasing Pt capping layer thickness up to a broad peak followed by a decrease to constant value with further increases in thickness. The possible intrinsic contributions to the enhancement of the damping in region I can be attributed to *d-d* hybridization and spin pumping in the case of the Pt capping layer. Hybridization causes changes in the electronic structure in the interface, whereas spin pumping allows absorption of angular momentum from the precessing magnetization, giving rise to an enhancement of damping. Both effects are intrinsic in nature. However, the effect of spin pumping should be very limited over the studied thickness range of Pt, as it is comparable or below the spin diffusion length. Hence, *d-d* hybridization mainly contributes to the intrinsic enhancement of the damping. Thus, a key mechanism for damping enhancement is the hybridization of 5*d* electrons of NM layer with the 3*d* electrons

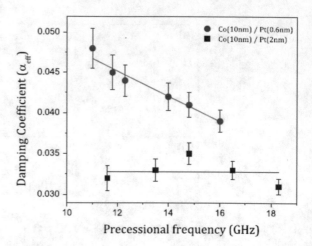

Fig. 7.4 Damping coefficient α_{eff} as a function of precessional frequency for 10-nm Co films capped with 0.6 and 2-nm Pt. The extrinsic damping decreases with the increasing Pt thickness until it reaches to negligible extrinsic effect at t_{NM} = 2-nm. *Reprinted with permission from Ref. [18]. Copyright 2016 by American Physical Society*

in the FM at the interface. This explanation is consistent with previous studies of FM-NM interfaces including the recent theoretical work of Barati et al. [19]. Extrinsic contributions to the damping can be attributed to two-magnon scattering, which is related to local variations linked to topological roughness, defects, and impurities at the interface [20–22]. For discontinuous Pt, it is suggested that variations in local electronic properties of the Pt capped and uncapped regions lead to local variation of intrinsic damping, which gives rise to extrinsic damping via two-magnon scattering. The precessional frequency dependence of the damping data (see Fig. 7.4) shows a linear increase in damping with the decreasing frequency for the discontinuous Pt layer, which clearly indicates that extrinsic effects are present in the system when the capping layer is discontinuous.

Following the initial rapid increase in region I, the damping reaches a broad peak in region II. From the structural analysis, it is observed that the second region falls into the thickness range where a continuous Pt capping layer is just forming. Beyond this thickness, a complete Pt layer is established. In region III, damping falls from its maximum and slowly stabilizes to an intermediate value with increasing Pt thickness. The independence of the damping on applied field indicates the mechanism here is predominantly intrinsic when the Pt layer is continuous. The decrease in the damping from the peak primarily represents a reduction in the extrinsic contribution that largely vanishes for higher thicknesses of Pt. It is interesting to observe that the final value of α_{eff} is larger than the uncapped (t_{NM} = 0) value for both Co and $Ni_{81}Fe_{19}$. The reason for this is that although at higher thicknesses of Pt the extrinsic contribution is negligible, intrinsic effects from the interface are the dominant contribution to the damping. Theoretical analysis shows some reduction in the intrinsic damping from the peak, along with periodic oscillations due to the formation of quantum well states. Oscillations are not observed in the experiment as the t_{NM} range is too small, but also any oscillations would be lost due to interfacial roughness.

Regarding the mechanism for extrinsic damping, it is suggested that two-magnon scattering is a result of local variation of d-d hybridization of the Co and Pt when the capping layer is discontinuous, leading to localized variations of the intrinsic damping. The miscibility of the NM material changes the local structure. In the case of Pt, this can produce Co-Pt clusters or islands at the surface, while for Au this leads to the formation of Au islands on Co in the low-thickness regime. For thin Pt, where the capping layer is incomplete, the formation of clusters or islands at the surface acts to break the translational symmetry leading to regions with higher and lower intrinsic damping due to the distribution of Co-Pt. The SOC for Pt is known to be stronger when it is 2D rather than 3D which increases the local damping of the Co-Pt islands. Thus, this inhomogeneous magnetic surface gives rise to an extrinsic contribution to the enhancement of the damping. This extrinsic contribution disappears once a continuous Pt layer forms over the Co layer.

It is worth noting that the experimental data (Fig. 7.2b) for the precessional frequency and the saturation magnetization, M_S, of the Co/Pt system show an increase in the net magnetic moment as the Pt capping increases. This may be attributed to a proximity-induced magnetization (PIM) in Pt [23, 24]. Due to partial oxidation of the uncapped Co surface, the M_S of uncapped Co is lower than expected for bulk Co. This is supported by the significant increase in M_S when t_{Pt} reaches 0.6 nm where the Pt cap would restrict the oxidation of the Co surface. Comparing the Pt and the Au layers, greater miscibility of Pt is an additional factor affecting the damping. As a result, two-magnon scattering is also lower in case of Co/Au. Furthermore, Au has a lower density of electrons in d-band states at the Fermi level compared to Pt, which therefore contributes very weakly to the intrinsic damping. For $Ni_{81}Fe_{19}$/Pt, the capping layer thickness dependence of the damping shows similar behavior to Co/Pt. However, it can be seen from Fig. 7.3a and c that α_{eff} for $Ni_{81}Fe_{19}$/Pt is higher at lower Pt thickness. This may be related to a slightly larger topological roughness and different FM crystal structure, as shown by the structural analysis. The larger enhancement of the damping for the $Ni_{81}Fe_{19}$, as compared to Co, may be explained by a higher spin-mixing conductance across the interface for fcc rather than hcp FM interfacial structure.

In Fig. 7.5, the structural changes associated with the thickness dependence of the damping are illustrated for regions, I, II, and III (cf. Fig. 7.5a) along with a comparison between the experimental damping data and the theoretical analysis of Co/Pt and Co/Au by Barati et al. (cf. Fig. 7.5b). The general trends of rapidly increasing damping to a peak followed by a small reduction and then leveling out of the damping with increasing film thickness are mostly similar as observed in the experiment and theory. Figure 7.5b shows slight reduction in the intrinsic damping from the peak. The spin-mixing conductance has been estimated for both Co/Pt and $Ni_{81}Fe_{19}$/Pt from the damping data obtained for samples with the thickest Pt layer, where Pt has formed a complete capping layer and measurements indicate that the mechanism for the enhanced damping is intrinsic. The spin-mixing conductance is related to the change in damping according to Eq. 7.8:

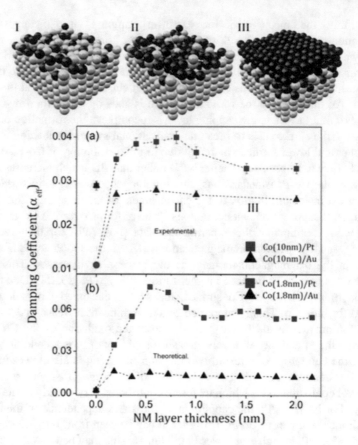

Fig. 7.5 Schematic illustration of the growth of discontinuous to continuous NM capping layer. **a** Experimental damping data for Co/Pt and Co/Au as a function of t_{NM}. The circular data point is a literature value for pure cobalt. **b** Theoretical variation in damping for Co/Pt and Co/Au adapted from Ref. [19] as a function of t_{NM}. *Reprinted with permission from Ref. [18]. Copyright 2016 by American Physical Society*

$$\Delta\alpha = \alpha_{\text{eff}} - \alpha_0 = \frac{g\mu_{\text{B}}}{4\pi t_{\text{FM}}M_{\text{eff}}} g_{\uparrow\downarrow}^{\text{eff}} \tag{7.8}$$

where g is the Land´e g-factor, μ_{B} is the Bohr magneton, M_{eff} is the effective magnetization, t_{FM} is the FM layer thickness and $g_{\uparrow\downarrow}^{\text{eff}}$ is the effective spin-mixing conductance.

Taking literature values for the damping in bulk *hcp* Co as 0.011 and *fcc* $Ni_{81}Fe_{19}$ as 0.01 and saturation magnetization as obtained experimentally, the effective spin-mixing conductance was estimated from the observed saturation enhancement to the damping. The obtained value of spin-mixing conductance for Co/Pt is ~38 nm^{-2} which is comparable with the value observed in experimental study. For $Ni_{81}Fe_{19}$/Pt, the spin-mixing conductance value is ~125 nm^{-2}. This is notably higher than the value obtained for Co/Pt, and it is suggested that this may be related to the different crystallographic structures at the interface with the Pt.

7.1.4 Local Modification of Interface Using Focused Ion Beam in NiFe/Pt, NiFe/Cr, and NiFe/Au

(a) NiFe/Pt

In general, the control of magnetic properties at the micro- and nanoscale is important for technological applications. The damping coefficient can be modified by introducing doping elements including rare earth or transition metal elements [25, 26], with the dopant introduced by co-deposition, usually co-sputtering. A disadvantage of this approach is that the entire material is doped. An alternative method for doping is direct irradiation with an ion beam of the desired dopant. In earlier studies, dopants including Cr, Tb, Gd, Ni, and Fe have been introduced as implanted ions. This approach can be used to introduce dopants into a localized area via lithography with an appropriate ion source. High-energy beams are needed to ensure adequate doping, requiring either a research accelerator or a commercially available accelerator-based ion implanter. High-energy heavy-ion irradiation is not usually compatible with local patterning. It is well established that the dopants introduced by ion irradiation affect the magnetic properties by introducing specific atomic species into the material, by altering the microstructure of the magnetic material (e.g., recrystallization or amorphization), or through intermixing in multilayered structures. Fassbender et al. in 2004 [27] and in 2006 [28] showed through their experimental work that the saturation magnetization, magnetic anisotropy, coercivity, and damping can be modified by direct ion implantation. Ion beam irradiation can also modify the magnetic behavior through direct modification of the structure of the material being irradiated, without the ions acting as a dopant. Ion bombardment of a solid interface results in thousands of intermixed atoms per single ion impact. Ar^+ ion irradiation has been shown to cause intermixing in Co/Pt layers, grain growth and increased interfacial roughening, changing the magnetization from perpendicular to in-plane, while intermixing in antiferromagnetic/ferromagnetic bilayers has been shown to modify the damping. Also, light ions, such as He^+, can cause intermixing, but have been shown to end up buried in the substrate.

In contrast to mask-based ion beam patterning, focused ion beam (FIB) irradiation allows direct nanoscale patterning by utilizing Ga^+ ions. In bilayered or multilayered FM/NM systems, low-dose FIB irradiation can be used to induce interfacial doping providing a route to locally modify the magnetic properties without substantial structural changes or damage. In 2015, Ganguly et al. [29] studied the sputter-deposited bilayer of NiFe(10 nm)/Pt(3 nm) on a thermally

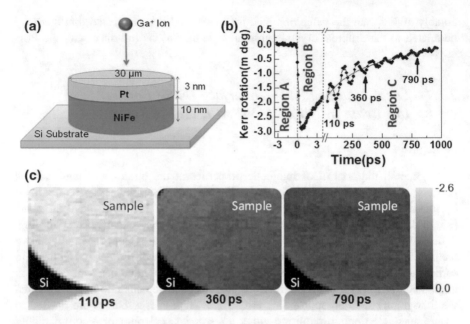

Fig. 7.6 **a** Schematic illustration of the ion beam irradiation of a bilayer circular structure. **b** TR-MOKE trace from a sample irradiated with ion dose $d = 3.1$ pC/μm^2. Note changes in the time base. **c** Time resolved Kerr images of the sample at three time delays. *Reprinted with permission from Ref.* [29]. *Open access material from the Nature Publishing Group*

oxidized Si[100] substrate. The bilayer films were patterned by electron beam lithography into 30-μm disks and were irradiated with irradiation dose (d) varying from 0 to 3.3 pC/μm^2 (cf. schematic Fig. 7.6a). Structurally, the as-deposited NiFe/Pt interface is known to have a typical width of less than 1 nm, resulting from a combination of topological roughness and chemical intermixing. An earlier detailed structural study of NiFe/Au revealed that with higher ion dose the interface rapidly becomes broader, the capping layer becomes very thin, and the layer develops into a compositionally graded alloy due to the intermixing of atoms of the heavy NM layer into the FM layer. Time-resolved magnetization dynamics of the samples were measured using a custom-built TR-MOKE microscope. Figure 7.6b shows typical time-resolved Kerr rotation data obtained from a bilayer disk irradiated with dose $d = 3.1$ pC/μm^2 measured under $H = 0.8$ kOe. In this, region A corresponds to negative delay, region B corresponds to ultrafast demagnetization ($\sim$500 fs), and region C shows the damped oscillatory magnetization signal superimposed on the decay signal. Scanning Kerr images of the sample were obtained by employing focused laser spot by using a piezoelectric scanning stage (x-y-z) with feedback loop for better stability. In Fig. 7.6c, the scanning Kerr images of the same region of one of the irradiated disks at three different antinodes of the precession (time delays = 110, 360, 790 ps as indicated by the arrows in Fig. 7.6b) are shown. These images show the uniformity of the excitation within the circular structures and decreasing brightness (Kerr angle) with increasing time delay as the precession is damped. From this, it can be inferred that there is almost negligible contribution to the damping from dephasing of multiple spin-wave modes and the damping is spatially uniform. Similar uniformity was observed for samples with other doses which confirmed that the magnetic changes due to irradiation were uniform over the sample and independent of the region probed. The effects of finite boundary or demagnetized region can also be excluded here since the sample size is much larger than the excitation and probed regions of about $1 \ \mu m^2$.

The influence of interfacial engineering on the precession and damping of NiFe/Pt bilayers is shown in Fig. 7.7a for three typical irradiation doses. For the lowest irradiation dose ($d = 0.3 \ pC/\mu m^2$), a value of $\alpha = 0.042$ was obtained, while for an intermediate dose, $d = 1.9 \ pC/\mu m^2$, α increased to 0.059 and subsequently decreased to $\alpha = 0.052$ for dose, $d = 3.3 \ pC/\mu m^2$. In Fig. 7.7b, the dependence of

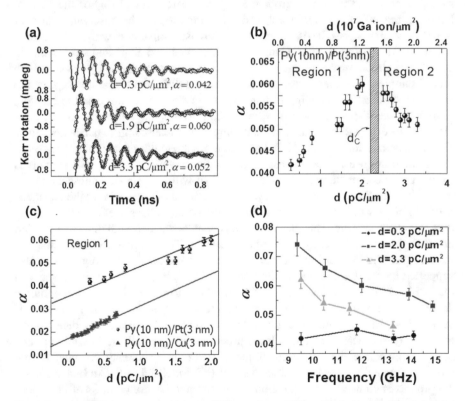

Fig. 7.7 a TR-MOKE traces at three different doses. Here, symbols correspond to the experimental data and solid curves are fits to damped sine curve. **b** Damping is plotted as a function of dose. The shaded box represents the transition between two regions. **c** Variation of damping with dose in the lower dose regime for NiFe/Pt (filled circles) and NiFe/Cu (filled triangles). Here, symbols are the experimental results and solid lines are linear fits. The data shown in (**a**–**c**) correspond to $H = 1.8$ kOe. **d** Variation of damping with precession frequency at three different doses. *Reprinted with permission from Ref.* [29]. *Open access material from the Nature Publishing Group*

damping on the ion beam dose is plotted, where the top x-axis shows the number density of Ga^+ ions used during irradiation. The variation in α can be divided into two distinct regions. In region 1, α increases monotonically with increasing d reaching a peak between 2.0 and 2.4 $pC/\mu m^2$. At higher doses, α decreases rapidly at first followed by a slower decrease with further increase in d. To understand the observed behavior, the variation in the lower dose regime (region 1) for NiFe/Pt was compared with that of NiFe(10 nm)/Cu(3 nm) bilayer in Fig. 7.7c. In both cases, α increased linearly with increasing d at similar rate of ~ 0.015 $\mu m^2/$ pC, but with a constant higher offset of 0.019 in the case of the NiFe/Pt bilayer. The origin of this shift is related to spin pumping and SOC associated with the Pt at the interface.

Earlier studies by Mizukami et al. have related the enhancement of α to spin pumping and SOC in the interfacial region and have described the Pt layer as a spin sink that absorbs spin waves from the adjacent FM layer [13]. For the case of NiFe/Cu, this effect is insignificant, and it is supported by the observation that for NiFe/Cu α approaches 0.015 at $d = 0$, which is comparable with the damping coefficient of uncapped NiFe. Ganguly et al. argued that SOC and spin pumping cannot explain the observed linear increase in α with dose that is common to both sets of samples. The doping from Ga is less than 1–2%, and this is not expected to significantly affect the damping of pure NiFe. However, with ion beam irradiation, NM atoms are displaced into the NiFe causing defects and structural changes that give rise to extrinsic two-magnon scattering, which causes an enhancement in α. The defects introduced in the NiFe layer increase the width of intermixed region as the irradiation dose increases, resulting in a linear variation of α with d. This is supported by earlier study by King et al. [30] (it will be discussed subsequently in this section) in case of NiFe/Au bilayers, where it was suggested that the enhanced α of NiFe/Au with irradiation was the result of an extrinsic effect attributed to two-magnon scattering. This mechanism has been associated with defects, disorder, or misfit symmetry breaking at the interface layer. The dominance of two-magnon scattering as the mechanism for enhanced α with increasing d in region 1 is also supported by the results from studies concerned with interface of an FM layer with Pt. In Fig. 7.7b, the variation of α with further increase of d shows a monotonic decrease in region 2. The shaded region indicates the dose region where α is expected to be maximum according to the trend of experimental data points. Structurally, ion beam irradiation leads predominantly to ballistic intermixing, and some insights into the evolution of the NiFe/Pt interface may be gained from previous detailed x-ray structural analysis of FIB irradiated NiFe/Au bilayers since the process is dominated by momentum transfer and the atomic mass of Pt is close to that of Au [31, 32]. The interface in as-deposited NiFe/Au and NiFe/Pt has some small intrinsic width (~ 1 nm from structural investigation). With increasing irradiation dose, the interface width increases linearly due to intermixing of the layers thus creating an increased thickness of compositionally graded NiFePt alloy between the NiFe and Pt. Previous experimental analysis of the Au capped system showed that a dose of roughly 1 $pC/\mu m^2$ creates a graded alloy interlayer of ~ 4 nm and sputter loss of more than 1 nm of NM capping layer. Extending this to the NiFe/Pt system, and assuming continued linear intermixing, for a dose above 2 $pC/\mu m^2$, this would lead to the formation of a compositionally graded alloy that extends through the NiFe layer thickness and the loss of most of the pure NM capping layer. Interestingly, the compositionally graded alloy region forms a continuous magnetic layer with the remaining NiFe part of the film. In addition, the loss of pure Pt layer thickness due to sputtering of the Pt-rich surface region leads to a reduction of the SOC and interlayer spin diffusion contributing to the intrinsic α which results in the decrease of effective α.

Ganguly et al. [29] further investigated in detail the variation of α with precession frequency f (cf. Fig. 7.7d) in detail. Interestingly, for minimum dose ($d = 0.3$ $pC/\mu m^2$), α is independent of f, which is a signature of intrinsic damping; however, for doses $d = 2.0$ and 3.3 $pC/\mu m^2$, α decreases steadily with f with similar slopes indicating an extrinsic contribution to the damping in both the cases. The extrinsic effect is dominant when H is weak (lower frequencies). At higher field, f is higher and the magnetization dynamics are strongly driven by the field itself, and hence, scattering is suppressed. From this, it was inferred that the enhancement in α with decreasing f is related to the presence of two-magnon scattering effect. From the difference in slope in $\alpha(f)$ between $d = 0.3$ and 2.0 $pC/\mu m^2$ curves in Fig. 7.7d, it is evident that the extrinsic damping develops with intermixing which explains the linear variation of $\alpha(d)$ in region 1 of Fig. 7.7b. The similar slope for $d = 2.0$ and 3.3 $pC/\mu m^2$ suggests that the extrinsic contribution continues in region 2. Further irradiation beyond a certain dose does not increase the number of scattering defects in the pure NiFe layer due to the formation of a thick interface region.

In Fig. 7.8a, the FFT power spectra of the TR-MOKE data for three different doses are reproduced. A clear decrease of f is observed with increasing dose. The variation of f with d is shown in Fig. 7.8b. While increasing the d from 0.3 to 2.0 $pC/\mu m^2$, a large decrease in $f \sim 1.3$ GHz occurs whereas a smaller decrease ~ 0.4 GHz occurs as d further increases from 2.0 to 3.3 $pC/\mu m^2$. Earlier studies have found a reduction of magnetic moment of Ni thin film with the concentration of Pt atoms. Hence, the decrease of f in region 1 is likely to be related to the reduction of moment due to the presence of Pt in the vicinity of Ni and Fe in the intermixed region. In region 2, a NiFe-Pt alloy is established throughout a large fraction of the original bilayer. The precessional frequency is almost constant with increasing dose, because the probability of sputtered Pt atoms to reach the pure NiFe layer by traversing through the thick interface region is small. In Fig. 7.8c, the relaxation times τ_1 and τ_2 as a function of d are shown. The relaxation times are obtained from the decaying part of the time-resolved Kerr rotation data as discussed earlier in Fig. 7.6b. As mentioned earlier, τ_1 is related to the transfer of electron and spin energy to the lattice, while τ_2 is related to the transfer of lattice energy to the substrate and surroundings. The energy transfer rate depends on the specific heat of lattice (S). Figure 7.8c shows both τ_1 and τ_2 increase slightly with d. In particular, the change is significant (larger than error) before and after $d = 2.0$ $pC/\mu m^2$ which is an indirect indication of a change in lattice configuration due to the formation of alloyed layer in region 2.

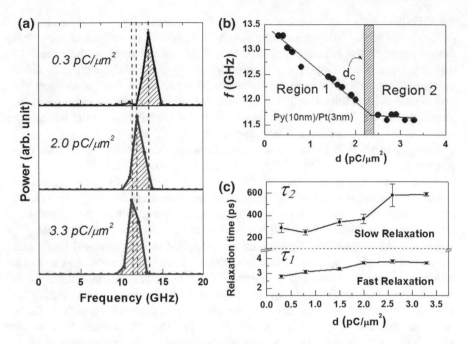

Fig. 7.8 a FFT power spectra of the TR-MOKE data of ion irradiated NiFe/Pt samples at three different doses at $H = 1.8$ kOe. **b** Variation of frequency f as a function of dose at $H = 1.8$ kOe. The *shaded box* represents the transition between two regions. **c** Magnetization relaxation times τ_1 and τ_2 are plotted as a function of dose. Here, the symbols are obtained from experimental data, while the *solid lines* are only guides to the eye. *Reprinted with permission from Ref.* [29]. *Open access material from the Nature Publishing Group*

(b) NiFe/Au and NiFe/Cr

In 2014, King et al. [30] investigated magnetization dynamics in thermally evaporated NiFe/Au and NiFe/Cr bilayer microstructures with 10 nm of NiFe thickness and 2–3 nm of either Au or Cr thickness using TR-MOKE. Arrays of circular structures of 30-µm diameter were patterned on the bilayer, and subsequently, FIB irradiation with different doses varying from 0 to 4 pC/µm^2 was performed. In Figs. 7.9 and 7.10, we reproduced the results of King et al. for NiFe/Au and NiFe/Cr bilayers, respectively. From Fig. 7.9a, we may notice that for unirradiated NiFe/Au, $\alpha \sim 0.01$, which is consistent with the typical values for $Ni_{80}Fe_{20}$ thin film. With increasing dose, α increases linearly with dose at a rate of 0.0035 µm^2/pC despite having some scatter in the data. From Fig. 7.9b, it may be seen that the precessional frequency is characterized by a general decrease with a small peak present between 1.3 and 2.0 pC/µm^2. For three different doses, the variation of damping with frequency is plotted in Fig. 7.9c. Similar plots for NiFe/Cr bilayer have been shown in Fig. 7.10a–c. As the damping parameter for the unirradiated NiFe/Au bilayer is comparable to NiFe, it was interpreted that the

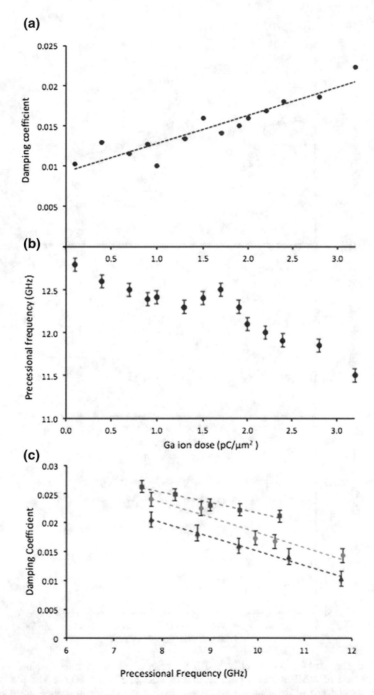

Fig. 7.9 a The damping coefficient, α, (note: error bars are smaller than data points) and **b** the precessional frequencies obtained from the TR-MOKE data of a NiFe/Au bilayer as a function of FIB dose at $H = 1.5$ kOe. **c** The precessional frequency dependence of the damping determined by varying H, for irradiation doses of 0.1 (triangle), 1.7 (circle), and 3.2 (square) pC/μm^2. *Reprinted with permission from Ref.* [30] *Copyright 2014 by the American Institute of Physics*

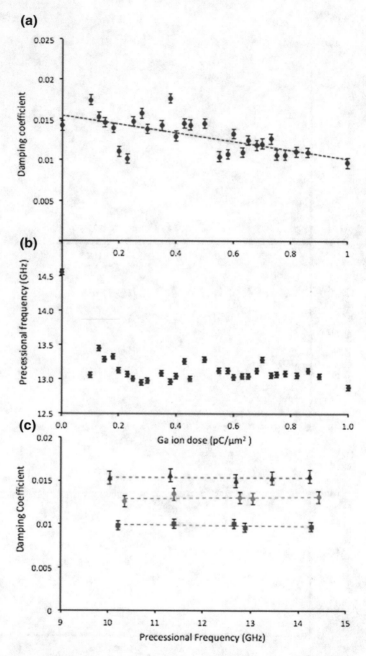

Fig. 7.10 **a** The damping coefficient, α, and **b** the precessional frequencies obtained from the TR-MOKE data of a NiFe/Cr as a function of FIB dose at $H = 1.5$ kOe. **c** The precessional frequency dependence of the damping determined by varying H, for irradiation doses of 0.13 (*triangle*), 0.40 (*circle*), and 1.00 (*square*) pC/μm^2. *Reprinted with permission from Ref.* [30]. *Copyright 2014 by the American Institute of Physics*

spin pumping is almost negligible. In this case, ion beam irradiation increases the precessional damping, which is associated with a broadening of the interfacial zone by intermixing between the NiFe and Au. This compositionally graded NiFeAu alloy extends over a few nanometers at the interface which increases the damping by enhanced scattering and/or modification of the spin–orbit interaction. The field dependent damping of the NiFe/Au was observed to decrease steadily with increasing frequency, indicating an extrinsic two-magnon type contribution to the damping that may be associated with increased disorder. Interestingly, the results reported by King et al. is in contrast to the behavior observed for molecular beam epitaxy (MBE) grown Au on Fe, where the damping was found to be enhanced by spin pumping in earlier studies [33]. Consistent with the proposal of enhanced scattering, increased electrical resistivity was observed for the NiFe/Au. Interestingly, the precessional frequency decreases with ion beam dose, but displays a small peak that is correlated in dose with a feature in the dose-dependent magnetic moment; however, the origin of this behavior remained unanswered in the study by King et al.

The use of Cr layer as capping on NiFe increases the value of α by ~50% compared to uncapped NiFe. With increasing ion dose, the damping coefficient reduces, falling to 0.0096 at the maximum dose, which is comparable with uncapped NiFe. The enhanced damping in comparison with NiFe could be an intrinsic effect associated with spin pumping across the interface or d-d hybridization of the Fe and Cr at the interface that increases the d-band width of the NiFe in contact with the Cr layer thus modifying the spin–orbit interaction.

It is also likely that the extrinsic two-magnon scattering linked to the coupling across the interface may increase the damping. However, the damping of NiFe/Cr remains invariant with increasing precessional frequency for the doses investigated (cf. Fig. 7.10b). This indicates an enhancement of the intrinsic damping, which contrasts with the effect of Cr on Fe observed for MBE grown samples that identified an extrinsic two-magnon scattering contribution to the damping. However, in the study of King et al. [30], the damping enhancement may be associated with the thickness of the Cr layer or the interfacial structure. Interestingly, both of these may be modified by ion-beam irradiation, with the loss of some Cr from the surface by sputtering and increased interfacial intermixing. This, in turn, may disrupt the spin pumping or interface hybridization, thereby reducing the damping, ultimately toward to a single NiFe layer value. The precessional frequency of the NiFe/Cr decreases sharply with low irradiation dose and subsequently shows small change for higher irradiation dose. King et al. argued that it is due to the interfacial hybridization that increases the NiFe moment and hence the precessional frequency of the unirradiated NiFe/Cr sample. Subsequent irradiation may degrade the magnetization, and the precessional frequency may decrease.

All the above-presented results and studies indicated that the combination of the low doses coupled with the high spatial resolution of the FIB may provide a methodology applicable for locally modifying the precessional magnetization behavior of ferromagnetic materials with feature sizes down to the nanoscale. This methodology may have applications to local control of damping for magnetic and spintronic device applications where static control of damping is expected to play crucial role.

7.2 Dynamic Control of Damping

One of the recent areas of modern spintronics research is to generate and utilize spin current for application in the devices. Liu et al. in 2011 [34] experimentally demonstrated the mechanism for controlling the magnetization dynamics in heavy metal/ferromagnet bilayer system by utilizing pure spin current generated by spin Hall effect (SHE) [35] that originates due to large spin–orbit coupling in heavy metal. This study triggered the idea of dynamical control of magnetic damping in such bilayers and multilayer system for practical application. As described in Chap. 6, the spin current due to SHE affects the magnetization dynamics of ferromagnetic layer placed next to the heavy metal layer. SHE applies a spin torque which is collinear with the damping torque, and the sign of it can be varied depending on the direction of flow of charge current through the heavy metal layer. Thus, charge current can be applied through the heavy metal layer, and one can evaluate the SHE-induced effect on magnetization dynamics via estimating modulation of effective damping in TR-MOKE experiments. An important parameter involved with the SHE is spin Hall angle which determines the conversion efficiency from charge current to spin current.

If a charge current is allowed to flow in the HM/FM (HM-heavy metals) bilayer sample, it gets distributed between HM and FM layers as determined by the resistivity of each layer. Due to strong spin–orbit interaction and impurity scattering within the HM, electrons with opposite spin polarity would be deflected in opposite directions, leading to spin accumulation at the interfaces and giving rise to a spin current due to the SHE. Spin current injected from the HM layer into the adjacent FM layer can exert a torque upon the magnetization which is referred to as the spin torque. This transfer of angular momentum to the magnetization modifies the magnetization relaxation τ and can thereby modify the effective magnetic damping α. It may also cause magnetization switching or setting auto oscillations of magnetization [36, 37]. The polarization of spin current, $\hat{\sigma}$, is determined by $\hat{J}_C \times \hat{n}$ where $\hat{J}_C$ is the applied charge current density and $\hat{n}$ is the normal vector to the interface plane. The interaction between spin current and magnetization has been described by the modified LLG equation:

$$\frac{\mathrm{d}\hat{m}}{\mathrm{d}t} = -\gamma(\hat{m} \times \vec{H}_{\mathrm{eff}}) + \alpha\left(\hat{m} \times \frac{\mathrm{d}\hat{m}}{\mathrm{d}t}\right) + \frac{\hbar}{2e\mu_0 M_S t_{\mathrm{FM}}} J_S(\hat{m} \times \hat{\sigma} \times \hat{m}) \qquad (7.9)$$

where the last term incorporates the anti-damping like spin–orbit torque (SOT) term. The direction of SOT is determined by $(\hat{m} \times \hat{\sigma} \times \hat{m})$ and acts collinearly with the effective damping, α, and can increase or decrease the effective value of α depending on the polarity of $\hat{\sigma}$. This modulation of the damping is related to the injected spin current density and the relative orientation of magnetic moment with respect to current by the equation:

$$\Delta\alpha = (\alpha - \alpha_0) = \hbar\gamma J_S \sin\theta/2eM_S t_{FM} 2\pi f, \qquad (7.10)$$

where α_0 is the original damping with zero dc current, e is the electronic charge, $\Delta\alpha$ is the change in the damping (MOD), and J_s is the spin-current density. The spin Hall angle, θ_{SH}, is given by J_S/J_C, defined as [34, 38, 39]:

$$J_S/J_C = 4\pi f e t_{FM} M_S \Delta\alpha/\hbar\gamma J_C \sin\theta \qquad (7.11)$$

Hence, by estimating the rate of MOD ($\Delta\alpha/J_C$) experimentally, an important parameter spin Hall angle given by J_S/J_C can be evaluated.

Spin current induced modulation of damping has been investigated by several research groups by electrical measurement techniques, in particular spin-torque ferromagnetic resonance (ST-FMR). Experiments on NiFe/Pt bilayer were performed by Liu et al. in 2011 [34] and subsequently by Ganguly et al. [38] and Kasai et al. both in 2014 [40]. In these studies, the estimated value of SHA of Pt was drastically different (by an order of magnitude) which led to intense debate over the experimental artifacts involved in the estimation of SHA. Subsequently in 2014, Ganguly et al. [39] proposed an unambiguous technique based on all-optical detection of magnetization dynamics using TR-MOKE to estimate the SHA via modulation of damping. The damping value in TR-MOKE experiments could be estimated directly in the time domain which is more advantageous than other techniques such as FMR linewidth measurement, where excitation of multiple modes may lead to inhomogeneous line broadening, which could artificially increase the damping. In the next two sub-sections, we discuss the modulation of damping in two exemplary systems along with spin Hall angle estimation in two different heavy metal layers.

7.2.1 Spin Hall Effect Induced Modulation of Damping in Pt/NiFe and W/CoFeB

Ganguly et al. in 2014 [39] used sputter-deposited thin film stacks of Pt(6.8 nm)/ $Ni_{81}Fe_{19}$(12.7 nm)/MgO(2.4 nm) which were grown on a thermally oxidized Si [100] substrate. The MgO layer was used as an insulating protective layer. For the experiments, 15-nm-thick Pt contact pads were used for making electrical connections, and the sample dimension was 5×1.5 mm^2. The resistivities of the Pt and $Ni_{81}Fe_{19}$ were estimated to be 15.0 and 36.0 $\mu\Omega$-cm, respectively. During the experiment, dc charge current (I_{dc}) along the length of the sample using a variable dc current source was applied. The experimental arrangement allowed simultaneous application of in-plane bias magnetic field with provision of varying the applied in-plane field angle. Due to negligibly small magneto-crystalline anisotropy of $Ni_{81}Fe_{19}$ thin film, the magnetization M was uniformly aligned along the direction of the bias magnetic field. Figure 7.11 shows a schematic of the sample and measurement arrangement. A 400 nm laser with 17.0 mJ/cm^2 of fluence and ~ 100 fs pulse width was used to excite the magnetization dynamics in the sample while a 800 nm laser with 2.1 mJ/cm^2 of fluence and ~ 80 fs pulse width probed the sample at different time delays with respect to the excitation.

Before investigating the effect of spin current on the magnetization dynamics, the sample was investigated thoroughly in the absence of any applied current. In this measurement, the damping for Pt/NiFe bilayer was estimated to be 0.021 and this value was independent of precession frequency, thus indicating absence of any extrinsic effect in the sample. The enhanced value of α for Pt/$Ni_{81}Fe_{19}$ bilayer stack in comparison with a single $Ni_{81}Fe_{19}$ layer may be explained by considering additional loss due to spin pumping into the Pt layer, as well as due to the d-d hybridization at the Pt/$Ni_{81}Fe_{19}$ interface. In order to investigate the effect of spin current on magnetization dynamics, dc current was subsequently applied to the bilayer stack. In Fig. 7.12a, the time-resolved magnetization precession at different current densities (J_c-current density through the Pt layer) varying from -1.1×10^{10} A/m^2 to $+1.1 \times 10^{10}$ A/m^2 with a magnetic field is shown. The bias magnetic field H ≈ 1.2 kOe was applied at an angle $\theta = 90°$ with respect to the current flow direction. A variation in α of up to $\pm 7\%$ as compared to its intrinsic value was observed for positive and negative values of J_c up to 1.1×10^{10} A/m^2. Significantly, for positive J_c, the damping α decreases whereas for negative J_c, α increases. The observed sensitivity of MOD on the sign of J_c suggests that the origin is related to the injected spin current due to SHE from the Pt layer, which depends on the current polarity. It is important to mention here that the Joule heating produced by the application of a dc current would only induce an increase

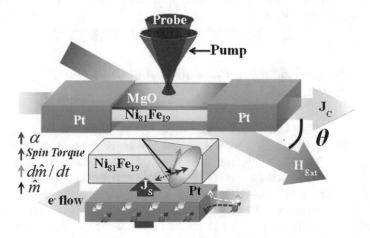

Fig. 7.11 Schematic diagram of the sample and experimental geometry. *Reprinted with permission from Ref.* [39]. *Copyright 2014 by the American Institute of Physics*

in α with respect to its zero current value and show no polarity dependence. Thus, the results suggest that Joule heating has a negligible influence on the magnetization dynamics in the Pt/Ni$_{81}$Fe$_{19}$ bilayer. In Fig. 7.12b, the variation of effective value of α with J_C is shown for $H \approx 1.2$ kOe where the field is orientated at $\theta = 0°$, $45°$, and $90°$ with respect to the current axis. At $\theta = 0°$, α remains almost constant for all J_c, indicating that there is no net torque on the magnetization. More interestingly, for $\theta = 45°$ and $90°$, a clear linear MOD is observed. The slope of the linear fit to the current density dependence of MOD determined for $\theta = 90°$ and $45°$ are 1.24×10^{-13} m^2/A and 8.14×10^{-14} m^2/A, indicating that the MOD depends not only on the polarity of J_C, but also on magnetization orientation with respect to the current. Figure 7.12c shows the variation of frequency as a function of J_c. A decrease (red shift) in the precessional frequency, f, with increasing magnitude of J_c was observed and was more or less symmetric with respect to zero current. Previously, a red shift in frequency with J_c was explained by considering the reduction in effective magnetization of the material because of magnetic fluctuation [2] due to Joule heating. However, the change in f with current should not affect α as is shown in Fig. 7.12c or the estimation of θ_{SH}.

More recently, using all-optical technique, Mondal et al., in 2017 [41], investigated the large spin Hall effect induced modulation of damping (MOD) in sputter-deposited substrate/W (t nm)/CoFeB (3 nm)/SiO$_2$ (2 nm) heterostructures (t − 2 to 7 nm) for varying W underlayer thickness. The interest in W has been primarily related to possible observation of large modulation of damping and thickness dependent phase transition (from β-phase to α-phase) exhibited by W in few nanometer film thickness regime.

7.2.2 Spin Hall Angle Estimation of Pt and W

From the results discussed in the previous subsection, an extremely important parameter, namely 'spin Hall angle (θ_{SH}),' which defines the conversion efficiency from charge current to spin current, can be estimated. By extracting the values of $\Delta\alpha/J_C$ obtained from Fig. 7.12b for $\theta = 45°$ and $90°$ and using in Eq. 7.11, θ_{SH} for Pt was estimated to be 0.11 ± 0.02 and 0.11 ± 0.03, respectively. From these values, an average value for θ_{SH} 0.11 ± 0.03 is calculated which is within the upper bound (<0.16) value reported in the literature. Most importantly, the estimated value of SHA is unambiguous, and it does not involve any experimental artifacts associated with the measurement technique.

Following similar calculation, the SHA for W as a function of thickness 2 nm $\leq t \leq$ 7 nm in Sub/W(t)/Co$_{20}$Fe$_{60}$B$_{20}$(3 nm)/SiO$_2$(2 nm) has been estimated by Mondal et al. [41]. Interestingly, they observed a giant value of SHA as large as 0.4 ± 0.04 for $t = 3$ nm. The estimated large value of SHA in β-W films indicates that in these heterostructures all the mechanisms efficiently contribute for the large SHA as well as the W/CoFeB interface is highly transparent. The trend found in the variation of SHA with W thickness (β-W has larger SHA in

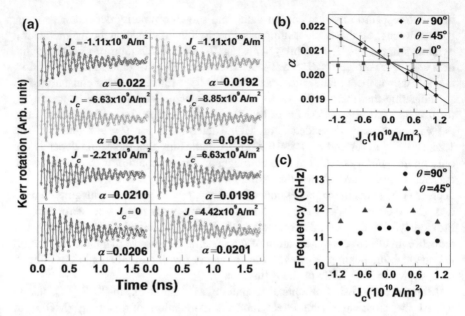

Fig. 7.12 a Time-resolved magnetization precession at different dc charge current densities, J_c, for $\theta = 90°$, obtained from the TR-MOKE experiment. Symbols represent the experimental data points, and *solid lines* are the theoretical fits. **b** The variation of α with J_c for magnetic fields oriented at angle, $\theta = 0°$, 45°, and 90° with respect to the current direction. **c** Variation of the precessional frequency f with J_c for $\theta = 45°$ and 90°. *Reprinted with permission from Ref.* [39]. *Copyright 2014 by the American Institute of Physics*

comparison with α-W) in their experiments is mostly consistent with earlier reported results.

It should be emphasized here that from the point of view of spintronic and magnonic device applications, the ability to dynamically and reversibly tune effective damping using external stimuli as charge current is of significant importance.

References

1. Hoffmann A, Bader SD (2015) Opportunities at the frontiers of spintronics. Phys Rev Appl 4 (4):047001. doi:10.1103/PhysRevApplied.4.047001
2. Kruglyak VV, Demokritov SO, Grundler D (2010) Magnonics. J Phys D Appl Phys 43 (26):260301. doi:10.1088/0022-3727/43/26/260301
3. Berger L (1996) Emission of spin waves by a magnetic multilayer traversed by a current. Phys Rev B 54(13):9353–9358. doi:10.1103/PhysRevB.54.9353
4. Brataas A, Tserkovnyak Y, Bauer GEW, Halperin BI (2002) Spin battery operated by ferromagnetic resonance. Phys Rev B 66(6):060404. doi:10.1103/PhysRevB.66.060404
5. Gilbert TL (2004) A phenomenological theory of damping in ferromagnetic materials. IEEE Trans Magn 40(6):3443–3449. doi:10.1109/tmag.2004.836740

6. Tserkovnyak Y, Brataas A, Bauer GEW (2002) Spin pumping and magnetization dynamics in metallic multilayers. Phys Rev B 66(22):224403. doi:10.1103/PhysRevB.66.224403

7. Mizukami S, Ando Y, Miyazaki T (2001) Ferromagnetic resonance linewidth for NM/80NiFe/NM films (NM = Cu, Ta, Pd and Pt). J Magn Magn Mater 226–230, Part 2, 1640–1642. doi:10.1016/S0304-8853(00)01097-0

8. Mizukami S, Ando Y, Miyazaki T (2001) The study on ferromagnetic resonance linewidth for NM/80NiFe/NM (NM = Cu, Ta, Pd and Pt) films. Jpn J Appl Phys 40(2A):580–585. doi:10.1143/jjap.40.580

9. Mizukami S, Ando Y, Miyazaki T (2002) Effect of spin diffusion on Gilbert damping for a very thin permalloy layer in Cu/permalloy/Cu/Pt films. Phys Rev B 66(10):104413. doi:10.1103/PhysRevB.66.104413

10. Kamberský V (1976) On ferromagnetic resonance damping in metals. Czech J Phys B 26 (12):1366–1383. doi:10.1007/bf01587621

11. Nakajima N, Koide T, Shidara T, Miyauchi H, Fukutani H, Fujimori A, Iio K, Katayama T, Nývlt M, Suzuki Y (1998) Perpendicular magnetic anisotropy caused by interfacial hybridization via enhanced orbital moment in Co/Pt multilayers: magnetic circular x-ray dichroism study. Phys Rev Lett 81(23):5229–5232. doi:10.1103/PhysRevLett.81.5229

12. Pal S, Rana B, Hellwig O, Thomson T, Barman A (2011) Tunable magnonic frequency and damping in [Co/Pd]$_8$ multilayers with variable Co layer thickness. Appl Phys Lett 98 (8):082501. doi:10.1063/1.3559222

13. Mizukami S, Sajitha EP, Watanabe D, Wu F, Miyazaki T, Naganuma H, Oogane M, Ando Y (2010) Gilbert damping in perpendicularly magnetized Pt/Co/Pt films investigated by all-optical pump-probe technique. Appl Phys Lett 96(15):152502. doi:10.1063/1.3396983

14. Lagae L, Wirix-Speetjens R, Eyckmans W, Borghs S, De Boeck J (2005) Increased Gilbert damping in spin valves and magnetic tunnel junctions. J Magn Magn Mater 286:291–296. doi:10.1016/j.jmmm.2004.09.083

15. Rantschler JO, Maranville BB, Mallett JJ, Chen P, McMichael RD, Egelhoff WF (2005) Damping at normal metal/permalloy interfaces. IEEE Trans Magn 41(10):3523–3525. doi:10.1109/tmag.2005.854956

16. Gerrits T, Schneider ML, Silva TJ (2006) Enhanced ferromagnetic damping in Permalloy/Cu bilayers. J Appl Phys 99(2):023901. doi:10.1063/1.2159076

17. Marcham MK, Yu W, Keatley PS, Shelford LR, Shafer P, Cavill SA, Qing H, Neudert A, Childress JR, Katine JA, Arenholz E, Telling ND, Laan GVD, Hicken RJ (2013) Influence of a Dy overlayer on the precessional dynamics of a ferromagnetic thin film. Appl Phys Lett 102 (6):062418. doi:10.1063/1.4792740

18. Azzawi S, Ganguly A, Tokaç M, Rowan-Robinson RM, Sinha J, Hindmarch AT, Barman A, Atkinson D (2016) Evolution of damping in ferromagnetic/nonmagnetic thin film bilayers as a function of nonmagnetic layer thickness. Phys Rev B 93(5):054402. doi:10.1103/PhysRevB.93.054402. Publisher's note, Phys Rev B 93(21):219902. doi:10.1103/PhysRevB.93.219902

19. Barati E, Cinal M, Edwards DM, Umerski A (2014) Gilbert damping in magnetic layered systems. Phys Rev B **90**(1) (2014). doi:10.1103/PhysRevB.90.014420

20. Woltersdorf G, Heinrich B (2004) Two-magnon scattering in a self-assembled nanoscale network of misfit dislocations. Phys Rev B 69(18):184417. doi:10.1103/PhysRevB.69.184417

21. Heinrich B, Bland JAC (2005) Spin relaxation in magnetic metallic layers and multilayers. In: Bland JAC (ed) Ultrathin magnetic structures: fundamentals of nanomagnetism, vol 3. Springer, New York

22. Heinrich B, Urban R, Woltersdorf G (2002) Magnetic relaxations in metallic multilayers. IEEE Trans Magn 38(5):2496–2501. doi:10.1109/tmag.2002.801906

23. Geissler J, Goering E, Justen M, Weigand F, Schütz G, Langer J, Schmitz D, Maletta H, Mattheis R (2001) Pt magnetization profile in a Pt/Co bilayer studied by resonant magnetic x-ray reflectometry. Phys Rev B 65(2):020405. doi:10.1103/PhysRevB.65.020405

24. Suzuki M, Muraoka H, Inaba Y, Miyagawa H, Kawamura N, Shimatsu T, Maruyama H, Ishimatsu N, Isohama Y, Sonobe Y (2005) Depth profile of spin and orbital magnetic

moments in a subnanometer Pt film on Co. Phys Rev B 72(5):054430 doi: 10.1103/PhysRevB.72.054430

25. Bailey W, Kabos P, Mancoff F, Russek S (2001) Control of magnetization dynamics in $Ni_{81}Fe_{19}$ thin films through the use of rare-earth dopants. IEEE Trans Magn 37(4):1749–1754. doi:10.1109/20.950957

26. Fu Y, Sun L, Wang JS, Bai XJ, Kou ZX, Zhai Y, Du J, Wu J, Xu YB, Lu HX, Zhai HR (2009) Magnetic properties of $(Ni_{83}Fe_{17})_{1-x}Gd_x$ thin films with diluted Gd doping. IEEE Trans Magn 45(10):4004–4007. doi:10.1109/tmag.2009.2024164

27. Fassbender J, Ravelosona D, Samson Y (2004) Tailoring magnetism by light-ion irradiation. J Phys D Appl Phys 37(16):R179. doi:10.1088/0022-3727/37/16/R01

28. Fassbender J, McCord J (2006) Control of saturation magnetization, anisotropy, and damping due to Ni implantation in thin $Ni_{81}Fe_{19}$ layers. Appl Phys Lett 88(25):252501. doi:10.1063/1.2213948

29. Ganguly A, Azzawi S, Saha S, King JA, Rowan-Robinson RM, Hindmarch AT, Sinha J, Atkinson D, Barman A (2015) Tunable magnetization dynamics in interfacially modified $Ni_{81}Fe_{19}$/Pt bilayer thin film microstructures. Sci Rep 5:17596. doi:10.1038/srep17596

30. King JA, Ganguly A, Burn DM, Pal S, Sallabank EA, Hase TPA, Hindmarch AT, Barman A, Atkinson D (2014) Local control of magnetic damping in ferromagnetic/non-magnetic bilayers by interfacial intermixing induced by focused ion-beam irradiation. Appl Phys Lett 104(24):242410. doi:10.1063/1.4883860

31. Arac E, Burn DM, Eastwood DS, Hase TPA, Atkinson D (2012) Study of focused-ion-beam-induced structural and compositional modifications in nanoscale bilayer systems by combined grazing incidence x ray reflectivity and fluorescence. J Appl Phys 111 (4):044324. doi:10.1063/1.3689016

32. Burn DM, Hase TPA, Atkinson D (2014) Focused-ion-beam induced interfacial intermixing of magnetic bilayers for nanoscale control of magnetic properties. J Phys-Condens Matter 26 (23):236002. doi:10.1088/0953-8984/26/23/236002

33. Woltersdorf G, Buess M, Heinrich B, Back CH (2005) Time resolved magnetization dynamics of ultrathin Fe(001) films: spin-pumping and two-magnon scattering. Phys Rev Lett 95(3):037401. doi:10.1103/PhysRevLett.95.037401

34. Liu L, Moriyama T, Ralph DC, Buhrman RA (2011) Spin-torque ferromagnetic resonance induced by the spin Hall effect. Phys Rev Lett 106(3):036601. doi:10.1103/PhysRevLett.106.036601

35. Hirsch JE (1999) Spin Hall effect. Phys Rev Lett 83(9):1834–1837. doi:10.1103/PhysRevLett.83.1834

36. Demidov VE, Urazhdin S, Edwards ERJ, Stiles MD, McMichael RD, Demokritov SO (2011) Control of magnetic fluctuations by spin current. Phys Rev Lett 107(10):107204. doi:10.1103/PhysRevLett.107.107204

37. Demidov VE, Urazhdin S, Ulrichs H, Tiberkevich V, Slavin A, Baither D, Schmitz G, Demokritov SO (2012) Magnetic nano-oscillator driven by pure spin current. Nat Mater 11 (12):1028–1031. doi:10.1038/nmat3459

38. Ganguly A, Kondou K, Sukegawa H, Mitani S, Kasai S, Niimi Y, Otani Y, Barman A (2014) Thickness dependence of spin torque ferromagnetic resonance in $Co_{75}Fe_{25}$/Pt bilayer films. Appl Phys Lett 104(7):072405. doi:10.1063/1.4865425

39. Ganguly A, Rowan-Robinson RM, Haldar A, Jaiswal S, Sinha J, Hindmarch AT, Atkinson DA, Barman A (2014) Time-domain detection of current controlled magnetization damping in Pt/$Ni_{81}Fe_{19}$ bilayer and determination of Pt spin Hall angle. Appl Phys Lett 105 (11):112409. doi:10.1063/1.4896277

40. Kasai S, Kondou K, Sukegawa H, Mitani S, Tsukagoshi K, Otani Y (2014) Modulation of effective damping constant using spin Hall effect. Appl Phys Lett 104(9):092408. doi:10.1063/1.4867649

41. Mondal S, Choudhury S, Jha N, Ganguly A, Sinha J, Barman A (2017) All-optical detection of spin Hall angle in W/CoFeB/SiO_2 heterostructures by varying thickness of the tungsten layer. Phys Rev B 96(5):054414. doi:10.1103/PhysRevB.96.054414

Chapter 8
Summary and Future Direction

Spin dynamics in ferromagnetic thin films, nanostructures, and heterostructures have drawn significant attention due to the application potential in various magnetic devices and the fundamental physics involved in it. In this book, we have described, in reasonable detail, the spin dynamics in various systems starting from its historical evolution. We reviewed experimental and theoretical results related to the ultrafast demagnetization, relaxation, magnetization precession, and magnetic damping in ferromagnetic metallic thin films, bilayers, and nanostructures with a particular focus on these systems when excited by femtosecond (fs) optical pulses. In order to better understand the experimental techniques, theoretical backgrounds of magnetization dynamics are discussed.

Specifically, we described the ultrafast demagnetization studies in Ni thin films where spin dynamics with characteristic time scales of few hundreds of fs have been observed [1]. Though some phenomenological and advanced models for explaining the experimental observations have been proposed, however, the complete understanding and interpretation of magnetization dynamics for ultrashort delays remain elusive. Subsequently, we discussed the Landau–Lifshitz–Gilbert equation, commonly used for phenomenological description of the precessional dynamics [2]. Various mechanisms involved in explaining the origin of Gilbert damping parameter have been presented from an experimental perspective without going much deep into the theoretical calculation. We chose a specific example available in the literature which verifies all principal features contained in the theory of two-magnon scattering as well as describes it as a source of extrinsic contribution to damping in ultrathin ferromagnetic thin films [3].

In last 20 years, there have been significant developments in different measurement techniques for investigating the spin dynamics. We have presented in detail the basic concepts and working principles of time-resolved magneto-optical Kerr effect (TR-MOKE), Brillouin light scattering (BLS), and ferromagnetic resonance (FMR) along with their variants. Subtleties involved in estimation of damping using time domain and frequency domain experimental data are discussed [4, 5]. A brief discussion on advantages and disadvantages of various experimental

© Springer International Publishing AG 2018

A. Barman and J. Sinha, *Spin Dynamics and Damping in Ferromagnetic Thin Films and Nanostructures*, https://doi.org/10.1007/978-3-319-66296-1_8

methods for investigating spin dynamics is included. These have been followed by the material aspects, particularly interface effects, in the samples that influence the spin dynamics. An interesting study which summarizes the systematic dependence of spin polarization of demagnetization time has been presented [6]. Consequences arising from different anisotropy contributions, magnetic damping and magnetization precession have been discussed. Given the recent focus of interest of spin-tronics research community, we have discussed the control of magnetization dynamics by current and optical means as observed in some of the interesting experimental investigation. Briefly, we described the spin–orbit torques which may either originate from spin Hall effect or Rashba-like effect to control the magneti-zation dynamics efficiently [7].

With the advancement in the capability of growing high-quality thin-film heterostructures and nanofabrication techniques, experimental realization of various theoretically proposed effects have been possible in last few years. One of the important goals of this book was to summarize the status of the experimental progress related to static and dynamic control of magnetic damping in non-magnetic metal (NM)/ferromagnet (FM) bilayer systems [8–10]. Relevant ingredients in understanding the control of damping in NM/FM are the mechanisms of spin pumping and interfacial d-d hybridization which play crucial roles in influencing the damping parameter. Both these mechanisms are discussed in detail. These concepts have been used to describe the experimental results related to static control of magnetic damping in various bilayer systems [11]. At the end, we pro-vide a discussion of spin current-based dynamic control of effective damping in heavy metal/ferromagnet bilayer systems. Spin Hall effect-generated spin current in heavy metal layer causes linear modulation of damping in such bilayers. By implementing an all-optical-based TR-MOKE technique to investigate the linear modulation of damping, one can unambiguously determine the spin Hall angle of a heavy metal layer.

It is worth to mention here that in the field of spin dynamics, still a number of issues remain to be addressed. Emerging concepts such as control of domain wall motion by means of spin–orbit torques and optomagnetic switching have triggered intense research to precisely control magnetization dynamics at much faster time scale [12, 13]. Ultrafast demagnetization of ferromagnetic thin-film heterostructures in presence of antisymmetric exchange interaction is likely to show interesting results [14]. Further exploration of heterostructures of materials beyond the rela-tively well-known materials may lead to the observation of completely new phe-nomena. Using 2D materials such as graphene, MoS_2 and topological insulators as neighboring layer in the ferromagnetic thin film can allow investigation of ultrafast spin dynamics with the imprint of exotic effects originating from these materials. The role of intrinsic and extrinsic contributions to damping in nanoscale structures prepared on non-magnet/ferromagnet film geometry needs in-depth investigation and understanding. Interestingly, in these systems, the presence of the third layer allows one to visualize these stacks as thin-film heterostructures with structure inversion asymmetry. In last few years, there have been significant interest in materials with broken inversion symmetry and strong spin–orbit coupling. In such systems Dzyaloshinskii–Moriya interaction (DMI), which is an antisymmetric exchange interaction, can stabilize canted spins [15]. Of particular interest in such materials are magnetic skyrmions which are particle-like chiral spin textures that are topologically protected. Magnetic skyrmions can arrange spontaneously into lattices, and charge currents can displace them at remarkably low current densities. Under the influence of spin current generated by spin Hall effect, it will be really interesting to explore rich optically induced ultrafast skyrmion dynamics and its correlation with the fundamental aspect of magnetism. Efficient generation and utilization of spin current for controlling magnetization dynamics continue to be an active research area [16, 17]. A key to this is to obtain in-depth understanding of the influence of antidamping torque and field-like torque from various interfaces in ferromagnetic thin-film heterostructures. Development of new materials/interfaces with large spin Hall angle is another challenging issue.

From the fundamental point of view, the exploration of nonlinearities in magnetization dynamics is important to understand. In experiments related to spin-transfer torque-driven magnetization dynamics in spin torque nano-oscillators or in assemblies of nano-oscillators, the signature of complex dynamics including chaos has been occasionally reported. However, detailed understanding of it is still lacking. Devices like nanowire waveguides and magnonic crystals employing magnetic heterostructures are crucial for the development of new class of spin-based electronics. A detailed study of propagating spin waves using BLS and standing spin-wave modes using TR-MOKE technique is expected to provide intriguing insight for utilizing them in magnonic devices. In order to fit in various applications, it will be essential to extend the present state of optical control of magnetization dynamics toward smaller nanoscale dimensions. Specifically, for heat-assisted magnetic recording which is expected to push the data storage capacity to higher level, rigorous research is being carried out by various research groups.

In summary, extensive areas of investigation of spin dynamics over various time scales and length scales in tailored magnetic materials where interface effects are pronounced are continuing to show fascinating results. These will be the subject of future investigation both from fundamental interest as well as applications in technologies where the spin degree of freedom is required for implementation. We envision the possibility that research in the field of spin dynamics will enable us to achieve energy efficient and fast information processing devices in the future.

References

1. Beaurepaire E, Merle JC, Daunois A, Bigot JY (1996) Ultrafast spin dynamics in ferromagnetic nickel. Phys Rev Lett 76(22):4250–4253. doi:10.1103/PhysRevLett.76.4250
2. Gilbert TL (2004) A phenomenological theory of damping in ferromagnetic materials. IEEE Trans Magn 40(6):3443–3449. doi:10.1109/tmag.2004.836740

3. Woltersdorf G, Heinrich B (2004) Two-magnon scattering in a self-assembled nanoscale network of misfit dislocations. Phys Rev B 69(18):184417. doi:10.1103/PhysRevB.69. 184417

4. van Kampen M, Jozsa C, Kohlhepp JT, LeClair P, Lagae L, de Jonge WJM, Koopmans B (2002) All-optical probe of coherent spin waves. Phys Rev Lett 88(22):227201. doi:10.1103/ PhysRevLett.88.227201

5. Pal S, Rana B, Hellwig O, Thomson T, Barman A (2011) Tunable magnonic frequency and damping in [Co/Pd]$_8$ multilayers with variable Co layer thickness. Appl Phys Lett 98 (8):082501. doi:10.1063/1.3559222

6. Mann A, Walowski J, Münzenberg M, Maat S, Carey MJ, Childress JR, Mewes C, Ebke D, Drewello V, Reiss G, Thomas A (2012) Insights into ultrafast demagnetization in pseudogap half-metals. Phys Rev X 2(4):041008. doi:10.1103/PhysRevX.2.041008

7. Garello K, Miron IM, Avci CO, Freimuth F, Mokrousov Y, Bluegel S, Auffret S, Boulle O, Gaudin G, Gambardella P (2013) Symmetry and magnitude of spin-orbit torques in ferromagnetic heterostructures. Nat Nanotechnol 8(8):587–593. doi:10.1038/nnano.2013.145

8. Ganguly A, Azzawi S, Saha S, King JA, Rowan-Robinson RM, Hindmarch AT, Sinha J, Atkinson D, Barman A (2015) Tunable magnetization dynamics in interfacially modified Ni$_{81}$Fe$_{19}$/Pt bilayer thin film microstructures. Sci Rep 5:17596. doi:10.1038/srep17596

9. King JA, Ganguly A, Burn DM, Pal S, Sallabank EA, Hase TPA, Hindmarch AT, Barman A, Atkinson D (2014) Local control of magnetic damping in ferromagnetic/non-magnetic bilayers by interfacial intermixing induced by focused ion-beam irradiation. Appl Phys Lett 104(24):242410. doi:10.1063/1.4883860

10. Azzawi S, Ganguly A, Tokaç M, Rowan-Robinson RM, Sinha J, Hindmarch AT, Barman A, Atkinson D (2016) Evolution of damping in ferromagnetic/nonmagnetic thin film bilayers as a function of nonmagnetic layer thickness. Phys Rev B 93(5):054402. doi:10.1103/PhysRevB. 93.054402

11. Barati E, Cinal M, Edwards DM, Umerski A (2014) Gilbert damping in magnetic layered systems. Phys Rev B 90(1) doi:10.1103/PhysRevB.90.014420

12. Torrejon J, Kim J, Sinha J, Mitani S, Hayashi M, Yamanouchi M, Ohno H (2014) Interface control of the magnetic chirality in CoFeB/MgO heterostructures with heavy-metal underlayers. Nat. Commun. 5 (article number 4655). doi:10.1038/ncomms5655

13. Lambert CH, Mangin S, Varaprasad B, Takahashi YK, Hehn M, Cinchetti M, Malinowski G, Hono K, Fainman Y, Aeschlimann M, Fullerton EE (2014) All-optical control of ferromagnetic thin films and nanostructures. Science 345(6202):1337–1340. doi:10.1126/ science.1253493

14. Hoffmann A, Bader SD (2015) Opportunities at the frontiers of spintronics. Phys Rev Appl 4 (4):047001. doi:10.1103/PhysRevApplied.4.047001

15. Hellman F, Hoffmann A, Tserkovnyak Y, Beach G, Fullerton E, Leighton C, MacDonald A, Ralph D, Arena D, Durr H, Fischer P, Grollier J, Heremans J, Jungwirth T, Kimmel A, Koopmans B, Krivorotov I, May S, Petford-Long A, Rondinelli J, Samarth N, Schuller I, Slavin A, Stiles M, Tchernyshyov O, Thiaville A, Zink B (2017) Interface-induced phenomena in magnetism. Rev Mod Phys 89(2):025006. doi:10.1103/RevModPhys.89. 025006

16. Demidov VE, Urazhdin S, Edwards ERJ, Stiles MD, McMichael RD, Demokritov SO (2011) Control of magnetic fluctuations by spin current. Phys Rev Lett 107(10):107204. doi:10.1103/ PhysRevLett.107.107204

17. Demidov VE, Urazhdin S, Ulrichs H, Tiberkevich V, Slavin A, Baither D, Schmitz G, Demokritov SO (2012) Magnetic nano-oscillator driven by pure spin current. Nat Mater 11 (12):1028–1031. doi:10.1038/nmat3459